ISEE Lower-Level
Subject Test Mathematics

Student Practice Workbook

+ Two Full-Length ISEE Lower-Level Math Tests

Math Notion

www.MathNotion.com

ISEE Lower-Level Subject Test Mathematics

ISEE Lower-Level Subject Test Mathematics

ISEE Lower-Level Subject Test Mathematics

Published in the United State of America By

The Math Notion

Web: WWW.MathNotion.com

Email: info@Mathnotion.com

Copyright © 2021 by the Math Notion. All rights reserved. No part of this publication may be reproduced, stored in a retrieval system, or transmitted in any form or by any means, electronic, mechanical, photocopying, recording, scanning, or otherwise, except as permitted under Section 107 or 108 of the 1976 United States Copyright Ac, without permission of the author.

All inquiries should be addressed to the Math Notion.

ISBN: 978-1-63620-084-2

The Math Notion

Michael Smith has been a math instructor for over a decade now. He launched the Math Notion. Since 2006, we have devoted our time to both teaching and developing exceptional math learning materials. As a test prep company, we have worked with thousands of students. We have used the feedback of our students to develop a unique study program that can be used by students to drastically improve their math scores fast and effectively. We have more than a thousand Math learning books including:

- **SAT Math Prep**
- **ACT Math Prep**
- **SSAT/ISEE Math Prep**
- **Mathematics Prep Grade 3 to 8**
- **Common Core Math Prep**
- **many Math Education Workbooks, Study Guides, Practice and Exercise Books**

As an experienced Math test preparation company, we have helped many students raise their standardized test scores—and attend the colleges of their dreams: We tutor online and in person, we teach students in large groups, and we provide training materials and textbooks through our website and through Amazon.

You can contact us via email at:

info@Mathnotion.com

ISEE Lower-Level Subject Test Mathematics

Get the Targeted Practice You Need to Ace the ISEE Lower-Level Math Test!

ISEE Lower-Level Subject Test Mathematics includes helpful examples, easy-to-follow instructions, and plenty of math practice problems to assist students to master each concept, brush up their problem-solving skills, and create confidence.

The ISEE Lower-Level math practice book provides numerous opportunities to evaluate basic skills along with abundant remediation and intervention activities. It is a skill that permits you to quickly master intricate information and produce better leads in less time.

Students can boost their test-taking skills by taking the book's two practice ISEE Lower-Level Math exams. All test questions answered and explained in detail.

Important Features of the ISEE Lower-Level Math Book:

- A **complete review** of ISEE Lower-Level math test topics,
- Over 2,500 practice problems covering all topics tested,
- The most important concepts you need to know,
- Clear and concise, easy-to-follow sections,
- Well designed for enhanced learning and interest,
- Hands-on experience with all question types,
- **2 full-length practice tests** with detailed answer explanations,
- Cost-Effective Pricing,

Powerful math exercises to help you avoid traps and pacing yourself to beat the ISEE Lower-Level test. Students will gain valuable experience and raise their confidence by taking ISEE lower math practice tests, learning about test structure, and gaining a deeper understanding of what is tested on the ISEE Lower-Level math. If ever there was a book to respond to the pressure to increase students' test scores, this is it.

WWW.MathNotion.COM

… So Much More Online!

- ✓ FREE Math Lessons
- ✓ More Math Learning Books!
- ✓ Mathematics Worksheets
- ✓ Online Math Tutors

For a PDF Version of This Book

Please Visit WWW.MathNotion.com

ISEE Lower-Level Subject Test Mathematics

Contents

Chapter 1 : Place Values and Number Sense 11
    Place Values 12
    Comparing and Ordering Numbers 13
    Numbers in Word Form 14
    Roman Numerals 15
    Rounding Numbers 16
    Odd or Even 17
    Repeating Patterns 18
    Growing Patterns 19
    Patterns: Numbers 20
    Answers of Worksheets 21

Chapter 2 : Whole Number Operations 23
    Adding Whole Numbers 24
    Subtracting Whole Numbers 25
    Multiplying Whole Numbers 26
    Dividing Hundreds 27
    Long Division by Two Digits 28
    Division with Remainders 28
    Rounding Whole Numbers 29
    Whole Number Estimation 30
    Answers of Worksheets 31

Chapter 3 : Number Theory 33
    Factoring Numbers 34
    Prime Factorization 34
    Divisibility Rules 35
    Greatest Common Factor 36
    Least Common Multiple 36
    Answers of Worksheets 37

Chapter 4 : Fractions and Mixed Numbers 39
    Simplifying Fractions 40
    Like Denominators 41
    Compare Fractions with Like Denominators 43
    More Than Two Fractions with Like Denominators 44
    Unlike Denominators 45
    Ordering Fractions 47

ISEE Lower-Level Subject Test Mathematics

    Denominators of 10, 100, and 1000 48
    Fractions to Mixed Numbers 50
    Mixed Numbers to Fractions 51
    Add and Subtract Mixed Numbers 52
    Answers of Worksheets 53

Chapter 5 : Decimals 57
    Adding and Subtracting Decimals 58
    Multiplying and Dividing Decimals 59
    Rounding Decimals 60
    Comparing Decimals 61
    Answers of Worksheets 62

Chapter 6 : Ratios and Rates 63
    Simplifying Ratios 64
    Writing Ratios 64
    Create a Proportion 65
    Proportional Ratios 65
    Similar Figures 66
    Word Problems 67
    Answers of Worksheets 69

Chapter 7 : Measurement 71
    Reference Measurement Units 72
    Metric Length Units 73
    Customary Length Units 73
    Metric Capacity Units 74
    Customary Capacity Units 74
    Metric Weight and Mass Units 75
    Customary Weight and Mass Units 75
    Temperature Units 76
    Time 77
    Money Amounts 78
    Money: Word Problems 79
    Answers of Worksheets 80

Chapter 8 : Algebraic Thinking 82
    Finding Rules 83
    Algebraic Word Problems 84
    Evaluate Expressions 85
    Variables and Expressions 86
    Answers of Worksheets 87

WWW.MathNotion.Com

ISEE Lower-Level Subject Test Mathematics

Chapter 9 : Geometric 89
   Identifying Angles 90
   Estimate Angle Measurements 91
   Measure Angles with a Protractor 92
   Polygon Names 93
   Classify Triangles 94
   Parallel Sides in Quadrilaterals 95
   Identify Rectangles 96
   Perimeter: Find the Missing Side Lengths 97
   Perimeter and Area of Squares 98
   Perimeter and Area of rectangles 99
   Find the Area or Missing Side Length of a Rectangle 100
   Area and Perimeter: Word Problems 101
   Circumference, Diameter, and Radius 102
   Volume of Cubes and Rectangle Prisms 103
   Answers of Worksheets 104

Chapter 10 : Three-Dimensional Figures 106
   Identify Three–Dimensional Figures 107
   Count Vertices, Edges, and Faces 108
   Identify Faces of Three–Dimensional Figures 109
   Answers of Worksheets 110

Chapter 11 : Symmetry and Transformations 111
   Line Segments 112
   Identify Lines of Symmetry 113
   Count Lines of Symmetry 114
   Parallel, Perpendicular and Intersecting Lines 115
   Answers of Worksheets 116

Chapter 12 : Data Graphs, and Statistics 117
   Mean and Median 118
   Mode and Range 119
   Graph Points on a Coordinate Plane 120
   Bar Graph 121
   Tally and Pictographs 122
   Dot plots 123
   Line Graphs 124
   Stem–And–Leaf Plot 125
   Scatter Plots 126
   Probability Problems 127

WWW.MathNotion.Com

ISEE Lower-Level Subject Test Mathematics

Answers of Worksheets 128
Chapter 13 : ISEE Lower-Level Practice Tests 131
ISEE Lower-Level Practice Test Answer Sheets 133
ISEE Lower-Level Practice Test 1 135
   Quantitative Reasoning 135
   Mathematics Achievement 145
ISEE Lower-Level Practice Test 2 153
   Quantitative Reasoning 153
   Mathematics Achievement 161
Chapter 14 : Answers and Explanations 169
Answer Key 169
Practice Tests 1: Quantitative Reasoning 171
Practice Tests 1: Mathematics Achievement 176
Practice Tests 2: Quantitative Reasoning 181
Practice Tests 2: Mathematics Achievement 185

ISEE Lower-Level Subject Test Mathematics

Chapter 1 : Place Values and Number Sense

Topics that you'll learn in this chapter:

- ✓ Place Values,
- ✓ Compare and Ordering Numbers,
- ✓ Numbers in Word Form,
- ✓ Roman Numerals,
- ✓ Rounding Numbers,
- ✓ Odd or Even,
- ✓ Repeating Patterns,
- ✓ Growing Patterns,
- ✓ Patterns: Numbers,

ISEE Lower-Level Subject Test Mathematics

Place Values

✏ Write numbers in expanded form.

1) Sixty–two   ___ + ___

2) fifty–six   ___ + ___

3) thirty–one   ___ + ___

4) forty–five   ___ + ___

5) twenty-eight   ___ + ___

✏ Circle the correct choice.

6) The 6 in 56 is in the

        Ones place   tens place   hundreds place

7) The 2 in 25 is in the

        Ones place   tens place   hundreds place

8) The 9 in 918 is in the

        Ones place   tens place   hundreds place

9) The 3 in 537 is in the

        Ones place   tens place   hundreds place

10) The 9 in 289 is in the

        Ones place   tens place   hundreds place

ISEE Lower-Level Subject Test Mathematics

Comparing and Ordering Numbers

🖎 Use less than, equal to or greater than.

1) 31 _____ 33   8) 42 _____ 36

2) 57 _____ 49   9) 55 _____ 55

3) 92 _____ 88   10) 57 _____ 75

4) 76 _____ 67   11) 28 _____ 38

5) 43 _____ 43   12) 19 _____ 15

6) 54 _____ 46   13) 82 _____ 90

7) 97 _____ 88   14) 78 _____ 84

🖎 Order each set numbers from least to greatest.

15) – 18, – 22, 28, – 17, 4   ___, ___, ___, ___, ___, ___

16) 19, –36, 11, – 12, 5   ___, ___, ___, ___, ___, ___

17) 27, – 56, 20, 1, – 27   ___, ___, ___, ___, ___, ___

18) 26, – 96, 2, – 26, 87, –75   ___, ___, ___, ___, ___, ___

19) –10, –71, 70, –26, –59, –39   ___, ___, ___, ___, ___, ___

20) 88, 4, 38, 7, 78, 9   ___, ___, ___, ___, ___, ___

21) 84, 14, 24, 0, 35, 22   ___, ___, ___, ___, ___, ___

WWW.MathNotion.Com

ISEE Lower-Level Subject Test Mathematics

Numbers in Word Form

 Write each number in words.

1) 372 _____

2) 605 _____

3) 550 _____

4) 351 _____

5) 793 _____

6) 647 _____

7) 3,219 _____

8) 5,326 _____

9) 2,842 _____

10) 4,691_____

11) 5,531_____

12) 7,360_____

13) 2,532_____

14) 8,014_____

15) 11,242_____

ISEE Lower-Level Subject Test Mathematics

Roman Numerals

✏ Write in Romans numerals.

1	I	11	XI	21	XXI
2	II	12	XI	22	XXII
3	III	13	XI I	23	XXIII
4	IV	14	XIV	24	XXIV
5	V	15	XV	25	XXV
6	VI	16	XVI	26	XXVI
7	VII	17	XVII	27	XXVII
8	VIII	18	XVIII	28	XXVIII
9	IX	19	XIX	29	XXIX
10	X	20	XX	30	XXX

1) 11 _____   2) 21 _____

3) 24 _____   4) 16 _____

5) 27 _____   6) 29 _____

7) 12 _____   8) 28 _____

9) 15 _____   10) 20 _____

11) Add 16 + 14 and write in Roman numerals. _____

12) Subtract 34 − 5 and write in Roman numerals. _____

ISEE Lower-Level Subject Test Mathematics

Rounding Numbers

Round each number to the underlined place value.

1) 3,<u>7</u>93

2) 3,<u>8</u>76

3) 3,4<u>5</u>2

4) 7,1<u>9</u>3

5) 5,2<u>7</u>8

6) 1,4<u>7</u>7

7) 8,<u>3</u>13

8) 24.<u>6</u>8

9) 8<u>4</u>.92

10) 71.<u>3</u>4

11) 66<u>4</u>.7

12) <u>9</u>,135

13) 15.3<u>8</u>1

14) 4,<u>5</u>21

15) 3<u>6</u>.50

16) 4,<u>8</u>19

17) 6,6<u>8</u>5

18) 2,5<u>3</u>8

19) 73.<u>6</u>2

20) 16,<u>5</u>27

21) 2<u>9</u>.720

22) 12,3<u>6</u>6

23) 31,<u>7</u>29

24) 7,8<u>3</u>8

ISEE Lower-Level Subject Test Mathematics

Odd or Even

🖋 Identify whether each number is even or odd.

1) 18 _____

2) 27 _____

3) 21 _____

4) 17 _____

5) 67 _____

6) 76 _____

7) 80 _____

8) 53 _____

9) 58 _____

10) 98 _____

11) 49 _____

12) 113 _____

🖋 Circle the even number in each group.

13) 52, 11, 35, 73, 5, 29

14) 13, 15, 113, 87, 71, 18

15) 33, 45, 86, 59, 63, 87

16) 55, 32, 79, 51, 21, 83

🖋 Circle the odd number in each group.

17) 54, 36, 48, 76, 71, 100

18) 32, 56, 40, 74, 98, 67

19) 58, 92, 25, 78, 76, 50

20) 89, 12, 88, 42, 48, 120

WWW.MathNotion.Com

Repeating Patterns

🖉 Circle the picture that comes next in each picture pattern.

1)

 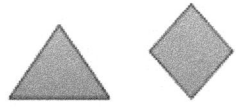

2)

3)

4)

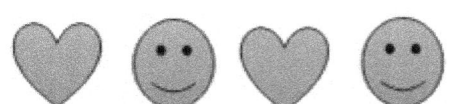 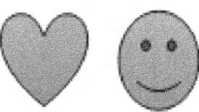

5)

ISEE Lower-Level Subject Test Mathematics

Growing Patterns

✏️ Draw the picture that comes next in each growing pattern.

1)

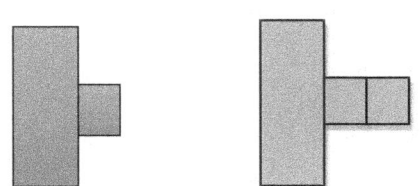

2)

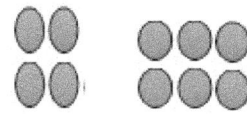

3)

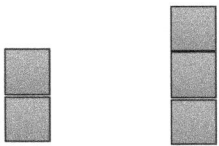

4)

5)

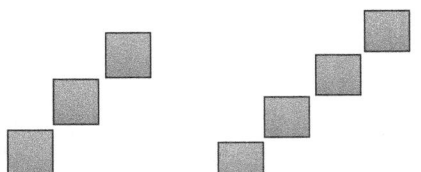

ISEE Lower-Level Subject Test Mathematics

Patterns: Numbers

✎ Write the numbers that come next.

1) 2, 5, 8, 11,   ____, ____, ____, ____

2) 10, 15, 20, 25,   ____, ____, ____, ____

3) 4, 8, 12, 16,   ____, ____, ____, ____

4) 7, 17, 27, 37,   ____, ____, ____, ____

5) 5, 12, 19, 26,   ____, ____, ____, ____

6) 8, 16, 24, 32, 40,   ____, ____, ____, ____

✎ Write the next three numbers in each counting sequence.

1) –31, –19, –7, ____, ____, ____, ____

2) 541, 526, 511, ____, ____, ____, ____

3) 14, 34, ____, ____, 94, ____

4) 21, 29, ____, ____, ____

5) 89, 78, ____, ____, ____

6) 95, 82, 69, ____, ____, ____

7) 198, 166, 134, ____, ____, ____

8) What are the next three numbers in this counting sequence?

   1870, 1970, 2070, ____, ____, ____

9) What is the fourth number in this counting sequence?

   8, 14, 20, ____

WWW.MathNotion.Com

ISEE Lower-Level Subject Test Mathematics

Answers of Worksheets

Place Values

1) 60 + 2
2) 50 + 6
3) 30 + 1
4) 40 + 5
5) 20 + 8
6) ones place
7) tens place
8) hundreds place
9) tens place
10) ones place

Comparing and Ordering Numbers

1) 31 less than 33
2) 57 greater than 49
3) 92 greater than 88
4) 76 greater than 67
5) 43 equals to 43
6) 54 greater than 46
7) 97 greater than 88
8) 42 greater than 36
9) 55 equals to 55
10) 57 less than 75
11) 28 less than 38
12) 19 greater than 15
13) 82 less than 90
14) 78 less than 84
15) −22, −18, −17, 4, 28
16) −36, −12, 5, 11, 19
17) −56, −27, 1, 20, 27
18) −96, −75, −26, 2, 26, 87
19) −71, −59, −39, −26, −10, 70
20) 4, 7, 9, 38, 78, 88
21) 0, 14, 22, 24, 35, 84

Numbers in Word Form

1) three hundred seventy-two
2) six hundred five
3) five hundred fifty
4) three hundred fifty-one
5) seven hundred ninety-three
6) six hundred forty-seven
7) three thousand, two hundred nineteen
8) five thousand, three hundred twenty-six
9) two thousand, eight hundred forty-two
10) four thousand, six hundred ninety-one
11) five thousand, five hundred thirty-one
12) seven thousand, three hundred sixty
13) two thousand, five hundred thirty-two
14) eight thousand, fourteen
15) eleven thousand, two hundred forty-two

Roman Numerals

1) XI
2) XXI
3) XXIV
4) XVI
5) XXVII
6) XXIX
7) XII
8) XXVIII
9) XV
10) XX
11) XXX
12) XXIX

Rounding Numbers

1) 4,000
2) 4,000
3) 3,450
4) 7,190
5) 5,280
6) 1,480
7) 8,300
8) 24.70
9) 85.00
10) 71.30
11) 665.00
12) 9,000

ISEE Lower-Level Subject Test Mathematics

13) 15.380
14) 4,500
15) 37.00
16) 4,800
17) 6,700
18) 2,540
19) 73.60
20) 16,500
21) 30.00
22) 12,370
23) 31,700
24) 7,840

Odd or Even

1) even
2) odd
3) odd
4) odd
5) odd
6) even
7) even
8) odd
9) even
10) even
11) odd
12) odd
13) 52
14) 18
15) 86
16) 32
17) 71
18) 67
19) 25
20) 89

Repeating pattern

1)
2)
3)
4)
5)

Growing patterns

1)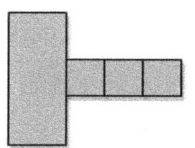
2)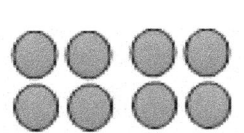
3)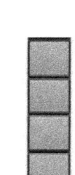
4)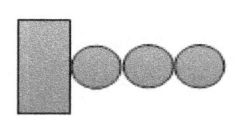
5)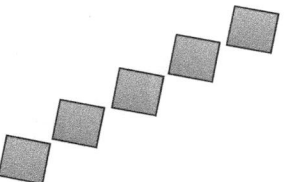

Patterns: Numbers

1) 2, 5, 8, 11, 14, 17, 20, 23
2) 10, 15, 20, 25, 30, 35, 40, 45
3) 4, 8, 12, 16, 20, 24, 28, 32
4) 7, 17, 27, 37, 47, 57, 67, 77
5) 5, 12, 19, 26, 33, 40, 47, 54
6) 8, 16, 24, 32, 40, 48, 56, 64

Patterns

1) 5, 17, 29, 41
2) 496, 481, 466, 451
3) 14, 34, 54, 74, 94, 114
4) 37, 45, 53
5) 67, 56, 45
6) 56, 43, 30
7) 102, 70, 38
8) 2170, 2270, 2370
9) 26

Chapter 2 : Whole Number Operations

Topics that you'll learn in this chapter:

- ✓ Adding Whole Numbers,
- ✓ Subtracting Whole Numbers,
- ✓ Multiplying Whole Numbers,
- ✓ Dividing Hundreds,
- ✓ Long Division by One Digit,
- ✓ Division with Remainders,
- ✓ Rounding Whole Numbers,
- ✓ Whole Number Estimation,

ISEE Lower-Level Subject Test Mathematics

Adding Whole Numbers

✎ Add.

1) 5,763
   + 8,238
   _____

2) 6,834
   + 4,998
   _____

3) 3,548
   + 5,693
   _____

4) 2,769
   + 8,872
   _____

5) 3,196
   + 2,936
   _____

6) 7,009
   + 4,992
   _____

✎ Find the missing numbers.

7) 3,468 + ___ = 4,102

8) 840 + 2,360 = ___

9) 5,200 + ___ = 7,980

10) 631 + ___ = 2,007

11) ___ + 803 = 3,945

12) ___ + 2,156 = 5,922

13) David sells gems. He finds a diamond in Istanbul and buys it for $4,795. Then, he flies to Cairo and purchases a bigger diamond for the bargain price of $9,633. How much does David spend on the two diamonds? _____

WWW.MathNotion.Com

ISEE Lower-Level Subject Test Mathematics

Subtracting Whole Numbers

✎ Subtract.

1) 10,512
   −4,411
   _____

2) 5,204
   −3,679
   _____

3) 8,520
   −6,483
   _____

4) 8,001
   −5,224
   _____

5) 11,916
   −8,711
   _____

6) 5,005
   −2,008
   _____

✎ Find the missing number.

7) 5,263 − ___ = 2,367

8) 7,198 − ___ = 4,742

9) 8,928 − 3,764 = ___

10) 6,511 − ___ = 3,759

11) 7,003 − 5,489 = ___

12) 8,800 − 5,995 = ___

13) Jackson had $7,189 invested in the stock market until he lost $3,793 on those investments. How much money does he have in the stock market now?

_____

WWW.MathNotion.Com

ISEE Lower-Level Subject Test Mathematics

Multiplying Whole Numbers

Find the answers.

1) 2,200 × 31

2) 3,200 × 22

3) 5,790 × 5

4) 5,220 × 3

5) 6,911 × 3

6) 1,998 × 40

7) 2,893 × 5.5

8) 2,254 × 3.5

9) 4,372 × 4.8

10) 3,984 × 2.75

11) 4,900 × 2.5

12) 8,200 × 4.5

WWW.MathNotion.Com

ISEE Lower-Level Subject Test Mathematics

Dividing Hundreds

✏ Find answers.

1) 4,440 ÷ 400

2) 1,600 ÷ 40

3) 9,990 ÷ 90

4) 4,200 ÷ 60

5) 6,400 ÷ 8,000

6) 2,700 ÷ 30

7) 3,333 ÷ 30

8) 558 ÷ 45

9) 2,278 ÷ 85

10) 1,683 ÷ 55

11) 1,582 ÷ 35

12) 9,000 ÷ 600

13) 1,000 ÷ 2,500

14) 44.8 ÷ 20

15) 6,800 ÷ 400

16) 1,500 ÷ 5,000

17) 36.60 ÷ 120

18) 7,700 ÷ 700

19) 5,400 ÷ 600

20) 8,000 ÷ 160

21) 18,000 ÷ 9,000

22) 42,000 ÷ 30

23) 480 ÷ 40

24) 63,000 ÷ 900

WWW.MathNotion.Com

ISEE Lower-Level Subject Test Mathematics

Long Division by Two Digits

✍ Find the quotient.

1) 18)576

2) 14)952

3) 21)588

4) 23)299

5) 44)748

6) 26)234

7) 16)496

8) 29)1,479

9) 54)1,080

10) 41)1,476

11) 53)2,491

12) 60)2,880

13) 32)2,912

14) 77)8,393

15) 85)3,740

16) 57)4,617

17) 50)9,200

18) 25)15,400

Division with Remainders

✍ Find the quotient with remainder.

1) 14)715

2) 16)2,750

3) 27)4,603

4) 58)2,554

5) 42)7,732

6) 63)6,737

7) 71)9,036

8) 65)8,624

9) 35)5,705

10) 92)13,161

11) 46)12,214

12) 69)42,482

13) 85)6,858

14) 87)34,304

ISEE Lower-Level Subject Test Mathematics

Rounding Whole Numbers

✎ Round each number to the underlined place value.

1) 7,<u>5</u>33

2) 9,<u>3</u>74

3) 8,8<u>8</u>3

4) 2,3<u>6</u>8

5) 5,5<u>7</u>7

6) 3,3<u>8</u>1

7) 3,<u>5</u>20

8) 9,3<u>3</u>8

9) 8.<u>5</u>81

10) 33.<u>5</u>7

11) 51.<u>6</u>9

12) 22.<u>1</u>38

13) <u>6</u>,758

14) 11,5<u>5</u>7

15) 8,8<u>3</u>8

16) 5.<u>8</u>89

17) 1.<u>8</u>60

18) 25.<u>0</u>70

19) <u>9</u>.332

20) 49.<u>4</u>8

21) 28.<u>8</u>9

22) 24,3<u>7</u>7

23) 52,1<u>5</u>8

24) 13,8<u>8</u>3

25) 9,<u>6</u>09

26) 17,4<u>5</u>1

27) 18,<u>7</u>68

ISEE Lower-Level Subject Test Mathematics

Whole Number Estimation

Estimate the sum by rounding each added to the nearest ten.

1) 875 + 325

2) 985 + 1,452

3) 2,424 + 4,128

4) 1,576 + 6,279

5) 1,247 + 3,863

6) 6,746 + 5,121

7) 3,924 + 6,456

8) 1,785 + 7,164

9) $1,458$
   $\underline{+2,442}$

10) $5,689$
    $\underline{+4,151}$

11) $8,259$
    $\underline{+4,754}$

12) $6,788$
    $\underline{+3,954}$

13) $9,123$
    $\underline{+4,455}$

14) $6,680$
    $\underline{+5,358}$

15) $3,165$
    $\underline{+7,124}$

16) $8,859$
    $\underline{+6,452}$

ISEE Lower-Level Subject Test Mathematics

Answers of Worksheets

Adding Whole Numbers

1) 14,001
2) 11,832
3) 9,241
4) 11,641
5) 6,132
6) 12,001
7) 634
8) 3,200
9) 2,780
10) 1,376
11) 3,142
12) 3,766
13) $14,428

Subtracting Whole Numbers

1) 6,101
2) 1,525
3) 2,037
4) 2,777
5) 3,205
6) 2,997
7) 2,896
8) 2,456
9) 5,164
10) 2,752
11) 1,514
12) 2,805
13) 3,396

Multiplying Whole Numbers

1) 68,200
2) 70,400
3) 28,950
4) 15,660
5) 20,733
6) 79,920
7) 15,911.5
8) 7,889
9) 20,985.6
10) 10,956
11) 12,250
12) 36,900

Dividing Hundreds

1) 11.1
2) 40
3) 111
4) 70
5) 0.8
6) 90
7) 111.1
8) 12.4
9) 26.8
10) 30.6
11) 45.2
12) 15
13) 0.4
14) 2.24
15) 17
16) 0.3
17) 0.305
18) 11
19) 9
20) 50
21) 2
22) 1,400
23) 12
24) 70

Long Division by Two Digits

1) 32
2) 68
3) 28
4) 13
5) 17
6) 9
7) 31
8) 51
9) 20
10) 36
11) 47
12) 48
13) 91
14) 109
15) 44
16) 81
17) 184
18) 616

WWW.MathNotion.Com

ISEE Lower-Level Subject Test Mathematics

Division with Remainders

1) 51 R1
2) 171 R14
3) 170 R13
4) 44 R2
5) 184 R4
6) 106 R59
7) 127 R19
8) 132 R44
9) 163 R0
10) 143 R5
11) 265 R24
12) 615 R47
13) 80 R58
14) 394 R26

Rounding Whole Numbers

1) 7,500
2) 9,400
3) 8,880
4) 2,370
5) 5,580
6) 3,380
7) 3,500
8) 9,340
9) 8.60
10) 33.60
11) 51.70
12) 22.100
13) 7,000
14) 11,560
15) 8,840
16) 5.900
17) 1.900
18) 25.100
19) 9.000
20) 49.50
21) 28.90
22) 24,380
23) 52,160
24) 13,880
25) 9,600
26) 17,450
27) 18,800

Whole Number Estimation

1) 1,200
2) 2,440
3) 6,550
4) 7,860
5) 5,110
6) 11,870
7) 10,380
8) 8,950
9) 3,900
10) 9,840
11) 13,010
12) 10,740
13) 13,580
14) 12,040
15) 10,290
16) 15,310

WWW.MathNotion.Com

Chapter 3 : Number Theory

Topics that you'll learn in this chapter:

- ✓ Factoring Numbers,
- ✓ Prime Factorization,
- ✓ Divisibility Rules,
- ✓ Greatest Common Factor,
- ✓ Least Common Multiple,

ISEE Lower-Level Subject Test Mathematics

Factoring Numbers

🔏 List all positive factors of each number.

1) 12   6) 56   11) 27
2) 16   7) 65   12) 63
3) 28   8) 70   13) 72
4) 34   9) 25   14) 15
5) 95   10) 48   15) 80

🔏 List the prime factorization for each number.

16) 10   19) 30   22) 55
17) 26   20) 40   23) 78
18) 20   21) 44   24) 96

Prime Factorization

🔏 Factor the following numbers to their prime factors.

1) 6   9) 58   17) 69
2) 49   10) 62   18) 76
3) 60   11) 75   19) 86
4) 4   12) 88   20) 92
5) 46   13) 93   21) 99
6) 57   14) 100   22) 77
7) 54   15) 68   23) 90
8) 38   16) 90   24) 74

WWW.MathNotion.Com

ISEE Lower-Level Subject Test Mathematics

Divisibility Rules

✎ Use the divisibility rules to underline the factors of the number.

1) 8   2 3 4 5 6 7 8 9 10

2) 18   2 3 4 5 6 7 8 9 10

3) 55   2 3 4 5 6 7 8 9 10

4) 45   2 3 4 5 6 7 8 9 10

5) 20   2 3 4 5 6 7 8 9 10

6) 9   2 3 4 5 6 7 8 9 10

7) 21   2 3 4 5 6 7 8 9 10

8) 28   2 3 4 5 6 7 8 9 10

9) 36   2 3 4 5 6 7 8 9 10

10) 40 2 3 4 5 6 7 8 9 10

11) 39 2 3 4 5 6 7 8 9 10

12) 51 2 3 4 5 6 7 8 9 10

ISEE Lower-Level Subject Test Mathematics

Greatest Common Factor

✏️ Find the GCF for each number pair.

1) 25, 15   9) 52, 3   17) 66, 18

2) 8, 18   10) 12, 54   18) 70, 15

3) 14, 28   11) 11, 13   19) 38, 14

4) 18, 32   12) 56, 48   20) 36, 28

5) 15, 45   13) 75, 25   21) 100, 60

6) 22, 33   14) 40, 60   22) 85, 35

7) 19, 21   15) 52, 32   23) 16, 48

8) 27, 72   16) 30, 55   24) 13, 39

Least Common Multiple

✏️ Find the LCM for each number pair.

1) 3, 15   9) 13, 26   17) 13, 2, 26

2) 5, 35   10) 15, 65   18) 18, 6, 24

3) 24, 16   11) 12, 8   19) 9, 12, 15

4) 28, 40   12) 6, 44   20) 7, 12, 4

5) 9, 27   13) 10, 16   21) 5, 15, 16

6) 46, 23   14) 7, 6   22) 13, 4, 26

7) 22, 66   15) 12, 36, 24 23) 3, 14, 5

8) 4, 9   16) 5, 11, 2   24) 32, 8, 3

WWW.MathNotion.Com

ISEE Lower-Level Subject Test Mathematics

Answers of Worksheets

Factoring Numbers

1) 1, 2, 3, 4, 6, 12
2) 1, 2, 4, 8, 16
3) 1, 2, 4, 7, 14, 28
4) 1, 2, 17, 34
5) 1, 5, 19, 95
6) 1, 2, 4, 7, 8, 14, 28, 56
7) 1, 5, 13, 65
8) 1, 2, 5, 7, 10, 14, 35, 70
9) 1, 5, 25
10) 1, 2, 3, 4, 6, 8, 12, 16, 24, 48
11) 1, 3, 9, 27
12) 1, 3, 7, 9, 21, 63
13) 1, 2, 3, 4, 6, 8, 9, 12, 18, 24, 36, 72
14) 1, 3, 5, 15
15) 1, 2, 4, 5, 8, 10, 16, 20, 40, 80
16) 2 × 5
17) 2 × 13
18) 2 × 2 × 5
19) 2 × 3 × 5
20) 2 × 2 × 2 × 5
21) 2 × 2 × 11
22) 5 × 11
23) 2 × 3 × 13
24) 2 × 2 × 2 × 2 × 2 × 3

Prime Factorization

1) 2. 3
2) 7. 7
3) 2. 2. 3. 5
4) 2. 2
5) 2. 23
6) 3. 19
7) 2. 3. 3. 3
8) 2. 19
9) 2. 29
10) 2. 31
11) 3. 5. 5
12) 2. 2. 2. 11
13) 3. 31
14) 2. 2. 5. 5
15) 2. 2. 17
16) 2. 3. 3. 5
17) 3. 23
18) 2. 2. 19
19) 2. 43
20) 2. 2. 23
21) 3. 3. 11
22) 7. 11
23) 2. 3. 3. 5
24) 2. 37

Divisibility Rules

1) 8   <u>2</u> 3 <u>4</u> 5 6 7 <u>8</u> 9 10
2) 18   <u>2</u> <u>3</u> 4 5 <u>6</u> 7 8 <u>9</u> 10
3) 55   2 3 4 <u>5</u> 6 7 8 9 10
4) 45   2 <u>3</u> 4 <u>5</u> 6 7 8 <u>9</u> 10
5) 20   <u>2</u> 3 <u>4</u> <u>5</u> 6 7 8 9 <u>10</u>
6) 9   2 <u>3</u> 4 5 6 7 8 <u>9</u> 10

WWW.MathNotion.Com

ISEE Lower-Level Subject Test Mathematics

7) 21   2 <u>3</u> 4 5 6 <u>7</u> 8 9 10

8) 28   <u>2</u> 3 <u>4</u> 5 6 <u>7</u> 8 9 10

9) 36   <u>2</u> <u>3</u> <u>4</u> 5 <u>6</u> 7 8 <u>9</u> 10

10) 40   <u>2</u> 3 <u>4</u> <u>5</u> 6 7 <u>8</u> 9 <u>10</u>

11) 39   2 <u>3</u> 4 5 6 7 8 9 10

12) 51   2 <u>3</u> 4 5 6 7 8 9 10

Greatest Common Factor

1) 5	7) 1	13) 25	19) 2
2) 2	8) 9	14) 20	20) 4
3) 14	9) 1	15) 4	21) 20
4) 2	10) 6	16) 5	22) 5
5) 15	11) 1	17) 6	23) 16
6) 11	12) 8	18) 5	24) 13

Least Common Multiple

1) 15	7) 66	13) 80	19) 180
2) 35	8) 36	14) 42	20) 84
3) 48	9) 26	15) 72	21) 240
4) 280	10) 195	16) 110	22) 52
5) 27	11) 24	17) 26	23) 210
6) 46	12) 132	18) 72	24) 96

ISEE Lower-Level Subject Test Mathematics

Chapter 4 : Fractions and Mixed Numbers

Topics that you'll learn in this chapter:

- ✓ Simplifying Fractions,
- ✓ Like Denominators,
- ✓ Compare Fractions with Like Denominators,
- ✓ More than two Fractions with Like Denominators,
- ✓ Unlike Denominators,
- ✓ Ordering Fractions,
- ✓ Denominators of 10, 100, and 1000,
- ✓ Fractions to Mixed Numbers,
- ✓ Mixed Numbers to Fractions,
- ✓ Add and Subtract Mixed Numbers,

ISEE Lower-Level Subject Test Mathematics

Simplifying Fractions

Simplify the fractions.

1) $\dfrac{44}{84}$

2) $\dfrac{8}{20}$

3) $\dfrac{12}{16}$

4) $\dfrac{4}{24}$

5) $\dfrac{15}{30}$

6) $\dfrac{9}{63}$

7) $\dfrac{4}{14}$

8) $\dfrac{17}{51}$

9) $\dfrac{24}{30}$

10) $\dfrac{5}{35}$

11) $\dfrac{16}{48}$

12) $\dfrac{33}{22}$

13) $\dfrac{45}{63}$

14) $\dfrac{2.4}{3.2}$

15) $\dfrac{12}{60}$

16) $\dfrac{70}{112}$

17) $\dfrac{2.7}{7.2}$

18) $\dfrac{33}{88}$

19) $\dfrac{1.5}{13.5}$

20) $\dfrac{39}{52}$

21) $\dfrac{5}{45}$

22) $\dfrac{2.1}{4.2}$

WWW.MathNotion.Com

ISEE Lower-Level Subject Test Mathematics

Like Denominators

✎ Add fractions.

1) $\dfrac{3}{4}+\dfrac{1}{4}$

2) $\dfrac{1}{5}+\dfrac{4}{5}$

3) $\dfrac{4}{9}+\dfrac{7}{9}$

4) $\dfrac{2}{7}+\dfrac{2}{7}$

5) $\dfrac{5}{13}+\dfrac{2}{13}$

6) $\dfrac{1}{14}+\dfrac{4}{14}$

7) $\dfrac{11}{19}+\dfrac{1}{19}$

8) $\dfrac{3}{16}+\dfrac{9}{16}$

9) $\dfrac{3}{10}+\dfrac{1}{10}$

10) $\dfrac{6}{17}+\dfrac{2}{17}$

11) $\dfrac{5}{22}+\dfrac{5}{22}$

12) $\dfrac{7}{35}+\dfrac{11}{35}$

13) $\dfrac{7}{27}+\dfrac{20}{27}$

14) $\dfrac{2}{31}+\dfrac{10}{31}$

15) $\dfrac{5}{23}+\dfrac{3}{23}$

16) $\dfrac{8}{41}+\dfrac{13}{41}$

17) $\dfrac{15}{37}+\dfrac{18}{37}$

18) $\dfrac{2}{51}+\dfrac{7}{51}$

19) $\dfrac{17}{26}+\dfrac{6}{26}$

20) $\dfrac{12}{48}+\dfrac{11}{48}$

21) $\dfrac{11}{29}+\dfrac{8}{29}$

22) $\dfrac{15}{34}+\dfrac{19}{34}$

23) $\dfrac{1}{19}+\dfrac{5}{19}$

24) $\dfrac{3}{53}+\dfrac{4}{53}$

25) $\dfrac{3}{20}+\dfrac{6}{20}$

26) $\dfrac{2}{63}+\dfrac{6}{63}$

27) $\dfrac{6}{38}+\dfrac{1}{38}$

28) $\dfrac{14}{31}+\dfrac{17}{31}$

29) $\dfrac{3}{28}+\dfrac{5}{28}$

30) $\dfrac{2}{37}+\dfrac{15}{37}$

WWW.MathNotion.Com

ISEE Lower-Level Subject Test Mathematics

✏ Subtract fractions.

1) $\dfrac{8}{9} - \dfrac{4}{9}$

2) $\dfrac{3}{8} - \dfrac{1}{8}$

3) $\dfrac{9}{11} - \dfrac{3}{11}$

4) $\dfrac{9}{14} - \dfrac{4}{14}$

5) $\dfrac{15}{20} - \dfrac{8}{20}$

6) $\dfrac{8}{15} - \dfrac{7}{15}$

7) $\dfrac{11}{19} - \dfrac{9}{19}$

8) $\dfrac{13}{16} - \dfrac{1}{16}$

9) $\dfrac{7}{29} - \dfrac{4}{29}$

10) $\dfrac{14}{23} - \dfrac{7}{23}$

11) $\dfrac{15}{34} - \dfrac{7}{34}$

12) $\dfrac{18}{41} - \dfrac{9}{41}$

13) $\dfrac{17}{39} - \dfrac{16}{39}$

14) $\dfrac{6}{26} - \dfrac{2}{26}$

15) $\dfrac{14}{17} - \dfrac{4}{17}$

16) $\dfrac{33}{55} - \dfrac{20}{55}$

17) $\dfrac{41}{49} - \dfrac{36}{49}$

18) $\dfrac{40}{53} - \dfrac{39}{53}$

19) $\dfrac{27}{37} - \dfrac{17}{37}$

20) $\dfrac{21}{47} - \dfrac{11}{47}$

21) $\dfrac{24}{43} - \dfrac{12}{43}$

22) $\dfrac{13}{19} - \dfrac{12}{19}$

23) $\dfrac{6}{26} - \dfrac{3}{26}$

24) $\dfrac{9}{15} - \dfrac{7}{15}$

25) $\dfrac{8}{39} - \dfrac{3}{39}$

26) $\dfrac{18}{61} - \dfrac{15}{61}$

27) $\dfrac{12}{53} - \dfrac{9}{53}$

28) $\dfrac{75}{76} - \dfrac{74}{76}$

29) $\dfrac{26}{45} - \dfrac{13}{45}$

30) $\dfrac{20}{57} - \dfrac{17}{57}$

WWW.MathNotion.Com

ISEE Lower-Level Subject Test Mathematics

Compare Fractions with Like Denominators

✎ Evaluate and compare. Write < or > or =.

1) $\frac{1}{3} + \frac{1}{3} \underline{\quad} \frac{1}{3}$

2) $\frac{3}{6} + \frac{3}{6} \underline{\quad} \frac{5}{6}$

3) $\frac{8}{9} - \frac{4}{9} \underline{\quad} \frac{7}{9}$

4) $\frac{4}{11} + \frac{5}{11} \underline{\quad} \frac{7}{11}$

5) $\frac{9}{14} - \frac{8}{14} \underline{\quad} \frac{5}{14}$

6) $\frac{11}{17} - \frac{3}{17} \underline{\quad} \frac{6}{17}$

7) $\frac{11}{21} + \frac{2}{21} \underline{\quad} \frac{10}{21}$

8) $\frac{8}{32} + \frac{6}{32} \underline{\quad} \frac{9}{32}$

9) $\frac{25}{29} - \frac{16}{29} \underline{\quad} \frac{11}{29}$

10) $\frac{28}{41} + \frac{13}{41} \underline{\quad} \frac{27}{41}$

11) $\frac{18}{35} - \frac{11}{35} \underline{\quad} \frac{22}{35}$

12) $\frac{32}{47} - \frac{22}{47} \underline{\quad} \frac{11}{47}$

13) $\frac{14}{27} + \frac{13}{27} \underline{\quad} \frac{24}{27}$

14) $\frac{34}{52} - \frac{11}{52} \underline{\quad} \frac{21}{52}$

15) $\frac{43}{56} - \frac{24}{56} \underline{\quad} \frac{27}{56}$

16) $\frac{27}{71} + \frac{25}{71} \underline{\quad} \frac{48}{71}$

WWW.MathNotion.Com

ISEE Lower-Level Subject Test Mathematics

More Than Two Fractions with Like Denominators

✎ Add fractions.

1) $\dfrac{5}{9} + \dfrac{2}{9} + \dfrac{2}{9}$

2) $\dfrac{4}{6} + \dfrac{1}{6} + \dfrac{1}{6}$

3) $\dfrac{2}{17} + \dfrac{4}{17} + \dfrac{2}{17}$

4) $\dfrac{1}{5} + \dfrac{1}{5} + \dfrac{1}{5}$

5) $\dfrac{7}{18} + \dfrac{2}{18} + \dfrac{3}{18}$

6) $\dfrac{3}{27} + \dfrac{5}{27} + \dfrac{2}{27}$

7) $\dfrac{4}{33} + \dfrac{4}{33} + \dfrac{4}{33}$

8) $\dfrac{8}{23} + \dfrac{6}{23} + \dfrac{2}{23}$

9) $\dfrac{13}{41} + \dfrac{2}{41} + \dfrac{8}{41}$

10) $\dfrac{6}{35} + \dfrac{9}{35} + \dfrac{20}{35}$

11) $\dfrac{1}{37} + \dfrac{5}{37} + \dfrac{5}{37}$

12) $\dfrac{4}{43} + \dfrac{9}{43} + \dfrac{8}{43}$

13) $\dfrac{4}{51} + \dfrac{10}{51} + \dfrac{7}{51}$

14) $\dfrac{5}{26} + \dfrac{13}{26} + \dfrac{6}{26}$

15) $\dfrac{5}{64} + \dfrac{4}{64} + \dfrac{2}{64}$

16) $\dfrac{1}{73} + \dfrac{5}{73} + \dfrac{6}{73}$

ISEE Lower-Level Subject Test Mathematics

Unlike Denominators

✏️ Add fraction.

1) $\dfrac{2}{9} + \dfrac{3}{4}$

2) $\dfrac{1}{4} + \dfrac{3}{5}$

3) $\dfrac{1}{16} + \dfrac{3}{4}$

4) $\dfrac{3}{8} + \dfrac{1}{7}$

5) $\dfrac{1}{3} + \dfrac{2}{4}$

6) $\dfrac{1}{6} + \dfrac{3}{7}$

7) $\dfrac{5}{18} + \dfrac{4}{6}$

8) $\dfrac{1}{12} + \dfrac{5}{6}$

9) $\dfrac{5}{27} + \dfrac{1}{9}$

10) $\dfrac{1}{6} + \dfrac{7}{24}$

11) $\dfrac{3}{5} + \dfrac{1}{8}$

12) $\dfrac{11}{42} + \dfrac{3}{7}$

13) $\dfrac{7}{20} + \dfrac{1}{3}$

14) $\dfrac{1}{45} + \dfrac{3}{5}$

15) $\dfrac{3}{32} + \dfrac{5}{8}$

16) $\dfrac{3}{48} + \dfrac{5}{6}$

17) $\dfrac{5}{12} + \dfrac{1}{6}$

18) $\dfrac{1}{34} + \dfrac{3}{17}$

19) $\dfrac{4}{9} + \dfrac{7}{54}$

20) $\dfrac{13}{56} + \dfrac{4}{7}$

21) $\dfrac{3}{12} + \dfrac{2}{3}$

22) $\dfrac{4}{33} + \dfrac{5}{11}$

WWW.MathNotion.Com

ISEE Lower-Level Subject Test Mathematics

✎ Subtract fractions.

1) $\dfrac{8}{9} - \dfrac{1}{2}$

2) $\dfrac{2}{3} - \dfrac{3}{10}$

3) $\dfrac{1}{6} - \dfrac{1}{9}$

4) $\dfrac{7}{8} - \dfrac{1}{4}$

5) $\dfrac{3}{4} - \dfrac{1}{28}$

6) $\dfrac{11}{30} - \dfrac{3}{15}$

7) $\dfrac{11}{18} - \dfrac{5}{9}$

8) $\dfrac{5}{13} - \dfrac{3}{26}$

9) $\dfrac{17}{35} - \dfrac{2}{7}$

10) $\dfrac{5}{6} - \dfrac{12}{36}$

11) $\dfrac{5}{9} - \dfrac{1}{27}$

12) $\dfrac{3}{5} - \dfrac{1}{8}$

13) $\dfrac{2}{3} - \dfrac{3}{5}$

14) $\dfrac{7}{8} - \dfrac{3}{7}$

15) $\dfrac{5}{9} - \dfrac{13}{45}$

16) $\dfrac{3}{4} - \dfrac{5}{36}$

17) $\dfrac{39}{49} - \dfrac{5}{7}$

18) $\dfrac{3}{11} - \dfrac{3}{22}$

19) $\dfrac{17}{48} - \dfrac{4}{12}$

20) $\dfrac{2}{3} - \dfrac{4}{13}$

21) $\dfrac{5}{8} - \dfrac{19}{72}$

22) $\dfrac{3}{5} - \dfrac{1}{12}$

WWW.MathNotion.Com

ISEE Lower-Level Subject Test Mathematics

Ordering Fractions

✎ Order the fractions from least to greatest.

1) $\frac{1}{5}, \frac{1}{11}, \frac{1}{8}, \frac{1}{3}$ ____, ____, ____, ____

2) $\frac{1}{9}, \frac{1}{18}, \frac{2}{4}, \frac{1}{5}$ ____, ____, ____, ____

3) $\frac{4}{7}, \frac{1}{7}, \frac{6}{21}, \frac{15}{21}$ ____, ____, ____, ____

4) $\frac{1}{2}, \frac{1}{3}, \frac{4}{9}, \frac{5}{18}$ ____, ____, ____, ____

5) $\frac{4}{9}, \frac{3}{4}, \frac{7}{36}, \frac{1}{6}$ ____, ____, ____, ____

✎ Order the fractions from greatest to least.

6) $\frac{3}{4}, \frac{4}{7}, \frac{3}{10}, \frac{5}{13}$ ____, ____, ____, ____

7) $\frac{5}{11}, \frac{5}{6}, \frac{2}{5}, \frac{1}{3}$ ____, ____, ____, ____

8) $\frac{7}{8}, \frac{1}{6}, \frac{3}{4}, \frac{5}{15}$ ____, ____, ____, ____

9) $\frac{4}{7}, \frac{2}{3}, \frac{11}{25}, \frac{13}{33}$ ____, ____, ____, ____

10) $\frac{18}{20}, \frac{15}{16}, \frac{14}{18}, \frac{5}{12}$ ____, ____, ____, ____

ISEE Lower-Level Subject Test Mathematics

Denominators of 10, 100, and 1000

✎ Add fractions.

1) $\dfrac{7}{10} + \dfrac{13}{100}$

2) $\dfrac{1}{10} + \dfrac{10}{100}$

3) $\dfrac{15}{100} + \dfrac{1}{1,000}$

4) $\dfrac{56}{100} + \dfrac{3}{10}$

5) $\dfrac{50}{1,000} + \dfrac{7}{10}$

6) $\dfrac{6}{10} + \dfrac{30}{1,000}$

7) $\dfrac{9}{100} + \dfrac{3}{10}$

8) $\dfrac{5}{10} + \dfrac{50}{100}$

9) $\dfrac{48}{100} + \dfrac{6}{10}$

10) $\dfrac{70}{100} + \dfrac{2}{10}$

11) $\dfrac{80}{100} + \dfrac{200}{1,000}$

12) $\dfrac{30}{100} + \dfrac{4}{10}$

13) $\dfrac{9}{100} + \dfrac{7}{10}$

14) $\dfrac{25}{100} + \dfrac{6}{10}$

15) $\dfrac{15}{100} + \dfrac{8}{10}$

16) $\dfrac{3}{10} + \dfrac{31}{100}$

17) $\dfrac{8}{10} + \dfrac{11}{100}$

18) $\dfrac{34}{100} + \dfrac{6}{10}$

WWW.MathNotion.Com

ISEE Lower-Level Subject Test Mathematics

✎ Subtract fractions.

1) $\dfrac{8}{10} - \dfrac{20}{100}$

2) $\dfrac{5}{10} - \dfrac{47}{100}$

3) $\dfrac{12}{100} - \dfrac{60}{1,000}$

4) $\dfrac{6}{10} - \dfrac{50}{100}$

5) $\dfrac{3}{10} - \dfrac{23}{100}$

6) $\dfrac{70}{100} - \dfrac{250}{1,000}$

7) $\dfrac{4}{10} - \dfrac{350}{1,000}$

8) $\dfrac{70}{100} - \dfrac{3}{10}$

9) $\dfrac{40}{100} - \dfrac{3}{10}$

10) $\dfrac{6}{10} - \dfrac{180}{1,000}$

11) $\dfrac{93}{100} - \dfrac{5}{10}$

12) $\dfrac{65}{100} - \dfrac{4}{10}$

13) $\dfrac{80}{100} - \dfrac{6}{10}$

14) $\dfrac{90}{100} - \dfrac{5}{10}$

15) $\dfrac{200}{1,000} - \dfrac{1}{10}$

16) $\dfrac{90}{100} - \dfrac{7}{10}$

17) $\dfrac{900}{1,000} - \dfrac{40}{100}$

18) $\dfrac{60}{100} - \dfrac{3}{10}$

WWW.MathNotion.Com

ISEE Lower-Level Subject Test Mathematics

Fractions to Mixed Numbers

✎ Convert fractions to mixed numbers.

1) $\dfrac{9}{5}$ 11) $\dfrac{41}{9}$

2) $\dfrac{11}{3}$ 12) $\dfrac{45}{12}$

3) $\dfrac{39}{8}$ 13) $\dfrac{17}{5}$

4) $\dfrac{27}{11}$ 14) $\dfrac{29}{6}$

5) $\dfrac{7}{2}$ 15) $\dfrac{13}{4}$

6) $\dfrac{43}{4}$ 16) $\dfrac{15}{7}$

7) $\dfrac{49}{9}$ 17) $\dfrac{65}{7}$

8) $\dfrac{15}{4}$ 18) $\dfrac{59}{8}$

9) $\dfrac{37}{7}$ 19) $\dfrac{25}{4}$

10) $\dfrac{19}{7}$ 20) $\dfrac{17}{8}$

WWW.MathNotion.Com

ISEE Lower-Level Subject Test Mathematics

Mixed Numbers to Fractions

✎ Convert to fraction.

1) $3\frac{3}{5}$

2) $1\frac{1}{3}$

3) $4\frac{2}{5}$

4) $4\frac{2}{8}$

5) $2\frac{1}{5}$

6) $2\frac{8}{11}$

7) $4\frac{4}{7}$

8) $3\frac{7}{12}$

9) $2\frac{1}{3}$

10) $7\frac{5}{7}$

11) $2\frac{7}{10}$

12) $3\frac{4}{9}$

13) $1\frac{5}{8}$

14) $4\frac{3}{11}$

15) $3\frac{4}{7}$

16) $5\frac{2}{8}$

17) $7\frac{1}{7}$

18) $13\frac{1}{2}$

19) $4\frac{2}{7}$

20) $5\frac{2}{10}$

21) $12\frac{1}{3}$

22) $7\frac{1}{8}$

WWW.MathNotion.Com

ISEE Lower-Level Subject Test Mathematics

Add and Subtract Mixed Numbers

🖎 Add mixed numbers.

1) $3\frac{2}{5} + 8\frac{1}{5}$

2) $3\frac{2}{3} + 4\frac{1}{2}$

3) $6\frac{2}{7} + 2\frac{3}{7}$

4) $4\frac{2}{5} + 3\frac{1}{4}$

5) $8\frac{3}{4} - 2\frac{1}{2}$

6) $6\frac{5}{12} - 4\frac{1}{4}$

7) $5\frac{3}{8} - 3\frac{7}{8}$

8) $6\frac{1}{4} - 2\frac{15}{16}$

9) $9\frac{23}{28} - 4\frac{17}{28}$

10) $6\frac{1}{6} + 6\frac{2}{3}$

11) $4\frac{2}{9} + 5\frac{5}{9}$

12) $2\frac{1}{4} + 7\frac{4}{7}$

13) $7\frac{1}{5} - 3\frac{3}{5}$

14) $3\frac{1}{6} + 2\frac{3}{7}$

15) $2\frac{1}{3} + 4\frac{1}{4}$

16) $4\frac{1}{4} - 1\frac{2}{5}$

17) $\frac{1}{3} + 6\frac{1}{6}$

18) $2\frac{3}{5} + 2\frac{1}{10}$

WWW.MathNotion.Com

ISEE Lower-Level Subject Test Mathematics

Answers of Worksheets

Simplifying Fractions

1) $\frac{11}{21}$
2) $\frac{2}{5}$
3) $\frac{3}{4}$
4) $\frac{1}{6}$
5) $\frac{1}{2}$
6) $\frac{1}{7}$
7) $\frac{2}{7}$
8) $\frac{1}{3}$
9) $\frac{4}{5}$
10) $\frac{1}{7}$
11) $\frac{1}{3}$
12) $\frac{3}{2}$
13) $\frac{5}{7}$
14) $\frac{3}{4}$
15) $\frac{1}{5}$
16) $\frac{5}{8}$
17) $\frac{3}{8}$
18) $\frac{3}{8}$
19) $\frac{1}{9}$
20) $\frac{3}{4}$
21) $\frac{1}{9}$
22) $\frac{1}{2}$

Like Denominators (addition)

1) 1
2) 1
3) $\frac{11}{9}$
4) $\frac{4}{7}$
5) $\frac{7}{13}$
6) $\frac{5}{14}$
7) $\frac{12}{19}$
8) $\frac{3}{4}$
9) $\frac{2}{5}$
10) $\frac{8}{17}$
11) $\frac{5}{11}$
12) $\frac{18}{35}$
13) 1
14) $\frac{12}{31}$
15) $\frac{8}{23}$
16) $\frac{21}{41}$
17) $\frac{33}{37}$
18) $\frac{3}{17}$
19) $\frac{23}{26}$
20) $\frac{23}{48}$
21) $\frac{19}{29}$
22) 1
23) $\frac{6}{19}$
24) $\frac{7}{53}$
25) $\frac{9}{20}$
26) $\frac{8}{63}$
27) $\frac{7}{38}$
28) 1
29) $\frac{2}{7}$
30) $\frac{17}{37}$

Like Denominators (Subtraction)

1) $\frac{4}{9}$
2) $\frac{1}{4}$
3) $\frac{6}{11}$
4) $\frac{5}{14}$
5) $\frac{7}{20}$
6) $\frac{1}{15}$
7) $\frac{2}{19}$
8) $\frac{3}{4}$
9) $\frac{3}{29}$
10) $\frac{7}{23}$
11) $\frac{8}{34}$
12) $\frac{9}{41}$
13) $\frac{1}{39}$
14) $\frac{2}{13}$
15) $\frac{10}{17}$
16) $\frac{13}{55}$
17) $\frac{5}{49}$
18) $\frac{1}{53}$
19) $\frac{10}{37}$
20) $\frac{10}{47}$
21) $\frac{12}{43}$
22) $\frac{1}{19}$
23) $\frac{3}{26}$
24) $\frac{2}{15}$

WWW.MathNotion.Com

ISEE Lower-Level Subject Test Mathematics

25) $\frac{5}{39}$ 27) $\frac{3}{53}$ 29) $\frac{13}{45}$

26) $\frac{3}{61}$ 28) $\frac{1}{76}$ 30) $\frac{1}{19}$

Compare Fractions with Like Denominators

1) $\frac{2}{3} > \frac{1}{3}$ 5) $\frac{1}{14} < \frac{5}{14}$ 9) $\frac{9}{29} < \frac{11}{29}$ 13) $1 > \frac{24}{27}$

2) $1 > \frac{5}{6}$ 6) $\frac{8}{17} > \frac{6}{17}$ 10) $1 > \frac{27}{41}$ 14) $\frac{23}{52} > \frac{21}{52}$

3) $\frac{4}{9} < \frac{7}{9}$ 7) $\frac{13}{21} > \frac{10}{21}$ 11) $\frac{7}{35} < \frac{22}{35}$ 15) $\frac{19}{56} < \frac{27}{56}$

4) $\frac{9}{11} > \frac{7}{11}$ 8) $\frac{14}{32} > \frac{9}{32}$ 12) $\frac{10}{47} < \frac{11}{47}$ 16) $\frac{52}{71} > \frac{48}{71}$

More Than Two Fractions with Like Denominators

1) 1   5) $\frac{2}{3}$ 9) $\frac{23}{41}$ 13) $\frac{7}{17}$

2) 1   6) $\frac{10}{27}$ 10) 1   14) $\frac{12}{13}$

3) $\frac{8}{17}$ 7) $\frac{4}{11}$ 11) $\frac{11}{37}$ 15) $\frac{11}{64}$

4) $\frac{3}{5}$ 8) $\frac{16}{23}$ 12) $\frac{21}{43}$ 16) $\frac{12}{73}$

Unlike Denominators (Addition)

1) $\frac{35}{36}$ 7) $\frac{17}{18}$ 13) $\frac{41}{60}$ 19) $\frac{31}{54}$

2) $\frac{17}{20}$ 8) $\frac{11}{12}$ 14) $\frac{28}{45}$ 20) $\frac{45}{56}$

3) $\frac{13}{16}$ 9) $\frac{8}{27}$ 15) $\frac{23}{32}$ 21) $\frac{11}{12}$

4) $\frac{29}{56}$ 10) $\frac{11}{24}$ 16) $\frac{43}{48}$ 22) $\frac{19}{33}$

5) $\frac{5}{6}$ 11) $\frac{29}{40}$ 17) $\frac{7}{12}$

6) $\frac{25}{42}$ 12) $\frac{29}{42}$ 18) $\frac{7}{34}$

Unlike Denominators (Subtraction)

1) $\frac{7}{18}$ 5) $\frac{5}{7}$ 9) $\frac{1}{5}$ 13) $\frac{1}{15}$

2) $\frac{11}{30}$ 6) $\frac{1}{6}$ 10) $\frac{1}{2}$ 14) $\frac{25}{56}$

3) $\frac{1}{18}$ 7) $\frac{1}{18}$ 11) $\frac{14}{27}$ 15) $\frac{4}{15}$

4) $\frac{5}{8}$ 8) $\frac{7}{26}$ 12) $\frac{19}{40}$ 16) $\frac{11}{18}$

WWW.MathNotion.Com

ISEE Lower-Level Subject Test Mathematics

17) $\frac{4}{49}$ 19) $\frac{1}{48}$ 21) $\frac{13}{36}$

18) $\frac{3}{22}$ 20) $\frac{14}{39}$ 22) $\frac{31}{60}$

Ordering Fractions

1) $\frac{1}{11}, \frac{1}{8}, \frac{1}{5}, \frac{1}{3}$ 5) $\frac{1}{6}, \frac{7}{36}, \frac{4}{9}, \frac{3}{4}$ 9) $\frac{2}{3}, \frac{4}{7}, \frac{11}{25}, \frac{13}{33}$

2) $\frac{1}{18}, \frac{1}{9}, \frac{1}{5}, \frac{2}{4}$ 6) $\frac{3}{4}, \frac{4}{7}, \frac{5}{13}, \frac{3}{10}$ 10) $\frac{15}{16}, \frac{18}{20}, \frac{14}{18}, \frac{5}{12}$

3) $\frac{1}{7}, \frac{6}{21}, \frac{4}{7}, \frac{15}{21}$ 7) $\frac{5}{6}, \frac{5}{11}, \frac{2}{5}, \frac{1}{3}$

4) $\frac{5}{18}, \frac{1}{3}, \frac{4}{9}, \frac{1}{2}$ 8) $\frac{7}{8}, \frac{3}{4}, \frac{5}{15}, \frac{1}{6}$

Denominators of 10, 100, and 1000

1) $\frac{83}{100}$ 6) $\frac{63}{100}$ 11) 1   16) $\frac{61}{100}$

2) $\frac{1}{5}$ 7) $\frac{39}{100}$ 12) $\frac{7}{10}$ 17) $\frac{91}{100}$

3) $\frac{151}{1,000}$ 8) 1   13) $\frac{79}{100}$ 18) $\frac{47}{50}$

4) $\frac{43}{50}$ 9) $\frac{27}{25}$ 14) $\frac{17}{20}$

5) $\frac{3}{4}$ 10) $\frac{9}{10}$ 15) $\frac{19}{20}$

Denominators of 10, 100, and 1000 (Subtract)

1) $\frac{3}{5}$ 6) $\frac{9}{20}$ 11) $\frac{43}{100}$ 16) $\frac{1}{5}$

2) $\frac{3}{100}$ 7) $\frac{1}{20}$ 12) $\frac{1}{4}$ 17) $\frac{1}{2}$

3) $\frac{3}{50}$ 8) $\frac{2}{5}$ 13) $\frac{1}{5}$ 18) $\frac{3}{10}$

4) $\frac{1}{10}$ 9) $\frac{1}{10}$ 14) $\frac{2}{5}$

5) $\frac{7}{100}$ 10) $\frac{21}{50}$ 15) $\frac{1}{10}$

Fractions to Mixed Numbers

1) $1\frac{4}{5}$ 5) $3\frac{1}{2}$ 9) $5\frac{2}{7}$ 13) $3\frac{2}{5}$

2) $3\frac{2}{3}$ 6) $10\frac{3}{4}$ 10) $2\frac{5}{7}$ 14) $4\frac{5}{6}$

3) $4\frac{7}{8}$ 7) $5\frac{4}{9}$ 11) $4\frac{5}{9}$ 15) $3\frac{1}{4}$

4) $2\frac{5}{11}$ 8) $3\frac{3}{4}$ 12) $3\frac{9}{12}$ 16) $2\frac{1}{7}$

WWW.MathNotion.Com

ISEE Lower-Level Subject Test Mathematics

17) $9\frac{2}{7}$ 18) $7\frac{3}{8}$ 19) $6\frac{1}{4}$ 20) $2\frac{1}{8}$

Mixed Numbers to Fractions

1) $\frac{18}{5}$ 7) $\frac{32}{7}$ 13) $\frac{13}{8}$ 19) $\frac{30}{7}$

2) $\frac{4}{3}$ 8) $\frac{43}{12}$ 14) $\frac{47}{11}$ 20) $\frac{52}{10}$

3) $\frac{22}{5}$ 9) $\frac{7}{3}$ 15) $\frac{25}{7}$ 21) $\frac{37}{3}$

4) $\frac{34}{8}$ 10) $\frac{54}{7}$ 16) $\frac{42}{8}$ 22) $\frac{57}{8}$

5) $\frac{11}{5}$ 11) $\frac{27}{10}$ 17) $\frac{50}{7}$

6) $\frac{30}{11}$ 12) $\frac{31}{9}$ 18) $\frac{27}{2}$

Add and Subtract Mixed Numbers

1) $11\frac{3}{5}$ 6) $2\frac{1}{6}$ 11) $9\frac{7}{9}$ 16) $2\frac{17}{20}$

2) $8\frac{1}{6}$ 7) $1\frac{1}{2}$ 12) $9\frac{23}{28}$ 17) $6\frac{1}{2}$

3) $8\frac{5}{7}$ 8) $3\frac{5}{16}$ 13) $3\frac{3}{5}$ 18) $4\frac{7}{10}$

4) $7\frac{13}{20}$ 9) $5\frac{3}{14}$ 14) $5\frac{25}{42}$

5) $6\frac{1}{4}$ 10) $12\frac{5}{6}$ 15) $6\frac{7}{12}$

Chapter 5 : Decimals

Topics that you'll learn in this chapter:

- ✓ Adding and Subtracting Decimals,
- ✓ Multiplying and Dividing Decimals,
- ✓ Round decimals,
- ✓ Comparing Decimals,

ISEE Lower-Level Subject Test Mathematics

Adding and Subtracting Decimals

🔹 Add and subtract decimals.

1) 24.19
   $-\underline{15.42}$
   $\underline{}$

2) 42.23
   $+\underline{25.42}$
   $\underline{}$

3) 72.54
   $+\underline{11.28}$
   $\underline{}$

4) 57.45
   $-\underline{24.75}$
   $\underline{}$

5) 43.57
   $+\underline{54.85}$
   $\underline{}$

6) 86.68
   $-\underline{54.12}$
   $\underline{}$

🔹 Solve.

7) ____ + 2.7 = 8.1

8) 6.4 + ____ = 12.8

9) 7.9 + ____ = 17

10) 4.6 + ____ = 15.3

11) ____ + 9.4 = 15

12) ____ + 8.24 = 13.54

🔹 Order each set of numbers from least to greatest.

1) 0.4, 0.67, 0.44, 0.73, 0.51 ___, ___, ___, ___, ___, ___

2) 3.9, 6.1, 4.28, 7.02, 4.65 ___, ___, ___, ___, ___, ___

3) 1.9, 1.04, 0.79, 0.72, 0.09 ___, ___, ___, ___, ___, ___

4) 2.6, 5.2, 1.9, 4.01, 1.99, 3.2 ___, ___, ___, ___, ___, ___

5) 4.2, 6.1, 3.8, 5.7, 2.1, 2.8 ___, ___, ___, ___, ___, ___

6) 0.56, 0.87, 0.14, 1.24, 3.1 ___, ___, ___, ___, ___, ___

WWW.MathNotion.Com

ISEE Lower-Level Subject Test Mathematics

Multiplying and Dividing Decimals

🖎 Find each product.

1) 1.5 × 2.1

2) 4.6 × 3.4

3) 6.3 × 2.5

4) 6.5 × 0.99

5) 12.1 × 5.2

6) 3.4 × 8.9

7) 4.8 × 9.1

8) 22.35 × 20

9) 15.25 × 3.6

🖎 Find each quotient.

10) 3.5 ÷ 0.85

11) 15.35 ÷ 4.6

12) 32.42 ÷ 8.8

13) 9.2 ÷ 3.4

14) 0.84 ÷ 0.1

15) 21.5 ÷ 1,000

16) 4.1 ÷ 100

17) 9.7 ÷ 10

18) 6.55 ÷ 1.25

19) 18.48 ÷ 11.2

WWW.MathNotion.Com

ISEE Lower-Level Subject Test Mathematics

Rounding Decimals

 Round each decimal number to the nearest place indicated.

1) 0.3<u>2</u>

2) 5.<u>0</u>1

3) 8.<u>8</u>24

4) 0.<u>4</u>78

5) <u>7</u>.32

6) 0.<u>2</u>9

7) 11.<u>3</u>1

8) <u>6</u>.223

9) 9.6<u>3</u>7

10) 5.<u>4</u>804

11) <u>7</u>.9

12) <u>5</u>.2439

13) 6.<u>4</u>92

14) 1.<u>6</u>2

15) 7<u>2</u>.85

16) 8<u>3</u>.67

17) 41.<u>6</u>8

18) 79<u>4</u>.741

19) 5<u>2</u>.2

20) 7<u>6</u>.93

21) <u>3</u>.219

22) 7<u>2</u>.09

23) 486.<u>4</u>91

24) 7.<u>0</u>8

ISEE Lower-Level Subject Test Mathematics

Comparing Decimals

✍ Write the correct comparison symbol (>, < or =).

1) 0.35 ___ 1.8

2) 1.9 ___ 1.19

3) 8.6 ___ 8.6

4) 2.45 ___ 24.5

5) 7.56 ___ 0.756

6) 11.4 ___ 11.05

7) 7.4 ___ 0.7.4

8) 8.56 ___ 0.85

9) 7 ___ 0.7

10) 7.12 ___ 0.712

11) 12.3 ___ 12.5

12) 4.67 ___ 4.68

13) 2.57 ___ 2.75

14) 3.46 ___ 0.346

15) 6.87 ___ 6.78

16) 0.89 ___ 0.98

17) 1.57 ___ 0.157

18) 0.092 ___ 0.091

19) 24.3 ___ 24.3

20) 0.17 ___ 0.71

21) 0.46 ___ 0.64

22) 0.2 ___ 0.08

23) 0.10 ___ 0.1

24) 3.52 ___ 31.5

WWW.MathNotion.Com

> ISEE Lower-Level Subject Test Mathematics

Answers of Worksheets

Adding and Subtracting Decimals

1) 8.77	4) 32.7	7) 5.4	10) 10.7
2) 67.65	5) 98.42	8) 6.4	11) 5.6
3) 83.82	6) 32.56	9) 9.1	12) 5.3

Order and Comparing Decimals

1) 0.4, 0.44, 0.51, 0.67, 0.73
2) 3.9, 4.28, 4.65, 6.1, 7.02
3) 0.09, 0.72, 0.79, 1.04, 1.9
4) 1.9, 1.99, 2.6, 3.2, 4.01, 5.2
5) 2.1, 2.8, 3.8, 4.2, 5.7, 6.1
6) 0.14, 0.56, 0.87, 1.24, 3.1

Multiplying and Dividing Decimals

1) 3.15	6) 30.26	11) 3.336…	16) 0.041
2) 15.64	7) 43.68	12) 3.684…	17) 0.97
3) 15.75	8) 447	13) 2.705…	18) 5.24
4) 6.435	9) 54.9	14) 8.4	19) 1.65
5) 62.92	10) 4.117…	15) 0.0215	

Rounding Decimals

1) 0.3	7) 11.3	13) 6.5	19) 52
2) 5.0	8) 6	14) 1.6	20) 77
3) 8.8	9) 9.64	15) 73	21) 3
4) 0.5	10) 5.5	16) 84	22) 72
5) 7	11) 8	17) 41.7	23) 486.5
6) 0.3	12) 5	18) 795	24) 7.1

Comparing Decimals

1) 0.35 < 1.8	9) 7 > 0.7	17) 1.57 > 0.157
2) 1.9 > 1.19	10) 7.12 > 0.712	18) 0.092 > 0.091
3) 8.6 = 8.6	11) 12.3 < 12.5	19) 24.3 = 24.3
4) 2.45 < 24.5	12) 4.67 < 4.68	20) 0.17 < 0.71
5) 7.56 > 0.756	13) 2.57 < 2.75	21) 0.46 < 0.64
6) 11.4 > 11.05	14) 3.46 > 0.346	22) 0.2 > 0.08
7) 7.4 > 0.74	15) 6.87 > 6.78	23) 0.10 = 0.1
8) 8.56 > 0.85	16) 0.89 < 0.98	24) 3.52 < 31.5

WWW.MathNotion.Com

Chapter 6 : Ratios and Rates

Topics that you'll learn in this chapter:

- ✓ Simplifying Ratios,
- ✓ Writing Ratios,
- ✓ Create a Proportion,
- ✓ Proportional Ratios,
- ✓ Similar Figures,
- ✓ Word Problems,

ISEE Lower-Level Subject Test Mathematics

Simplifying Ratios

Reduce each ratio.

1) 14: 56
2) 12: 36
3) 5: 35
4) 56: 48
5) 12: 14
6) 81: 63
7) 60: 3
8) 15: 10
9) 25: 20
10) 14: 28
11) 60: 84
12) 11: 99
13) 30: 45
14) 18: 45
15) 90: 15
16) 1.5: 3
17) 8: 88
18) 13: 52
19) 3: 75
20) 2.2: 22
21) 11: 33
22) 18: 81
23) 68: 80
24) 50: 500

Writing Ratios

Express each ratio as a rate and unite rate.

1) 180 miles on 6 gallons of gas.

2) 99 dollars for 11 books.

3) 35 miles on 3.5 gallons of gas

4) 7.5 inches of snow in 1.5 hours

Express each ratio as a fraction in the simplest form.

5) 6 feet out of 60 feet
6) 14 cakes out of 49 cakes
7) 32 dimes t0 60 dimes
8) 18 dimes out of 63 coins
9) 13 cups to 91 cups
10) 28 gallons to 42 gallons
11) 35 miles out of 120 miles
12) 22 blue cars out of 55 cars
13) 6.9 pennies to 69 pennies
14) 14 beetles out of 70 insects
15) 18 dimes to 54 dimes
16) 40 red cars out of 160 cars

ISEE Lower-Level Subject Test Mathematics

Create a Proportion

✎ Create proportion from the given set of numbers.

1) 1, 20, 4, 5

2) 9, 135, 1, 15

3) 5, 15, 8, 24

4) 49, 7, 4, 28

5) 9, 1, 108, 12

6) 45, 2, 5, 18

7) 28, 7, 24, 6

8) 11, 3, 55, 15

9) 5, 45, 36, 4

10) 16, 128, 1, 8

11) 2.5, 10, 5, 20

12) 5, 12, 15, 36

Proportional Ratios

✎ Solve each proportion.

1) $\frac{4}{8} = \frac{5}{d}$

2) $\frac{k}{6} = \frac{3}{9}$

3) $\frac{10}{8} = \frac{12}{x}$

4) $\frac{x}{15} = \frac{9}{5}$

5) $\frac{d}{11} = \frac{10}{1.1}$

6) $\frac{4.5}{6} = \frac{9}{x}$

7) $\frac{7}{15} = \frac{k}{60}$

8) $\frac{11}{1.5} = \frac{121}{d}$

9) $\frac{x}{0.7} = \frac{15}{2.8}$

10) $\frac{1.2}{4} = \frac{x}{2.5}$

11) $\frac{7.8}{x} = \frac{7.8}{2}$

12) $\frac{x}{3.4} = \frac{48}{16}$

13) $\frac{80}{20} = \frac{k}{60}$

14) $\frac{1.4}{5} = \frac{28}{d}$

15) $\frac{x}{7} = \frac{30}{15}$

16) $\frac{4}{1.6} = \frac{k}{1.6}$

17) $\frac{0.8}{1.2} = \frac{5.6}{d}$

18) $\frac{25}{x} = \frac{50}{4}$

19) $\frac{d}{9} = \frac{18}{27}$

20) $\frac{k}{12.6} = \frac{5}{12.6}$

21) $\frac{1.4}{7} = \frac{x}{10}$

22) $\frac{13}{5} = \frac{k}{15}$

23) $\frac{18}{21} = \frac{x}{7}$

24) $\frac{9}{99} = \frac{x}{22}$

WWW.MathNotion.Com

ISEE Lower-Level Subject Test Mathematics

Similar Figures

🖎 Each pair of figures is similar. Find the missing side.

1)

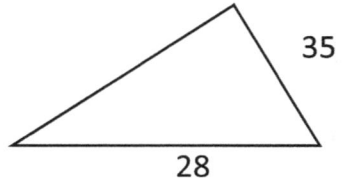

2)

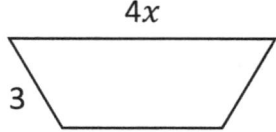

3)

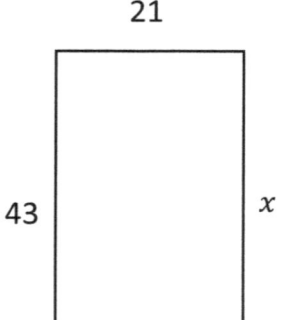

ISEE Lower-Level Subject Test Mathematics

Word Problems

Solve.

1) In a party, 15 soft drinks are required for every 18 guests. If there are 360 guests, how many soft drinks is required?

2) In Jack's class, 16 of the students are tall and 10 are short. In Michael's class 40 students are tall and 25 students are short. Which class has a higher ratio of tall to short students?

3) Are these ratios equivalent?
12 cards to 84 animals 18 marbles to 126 marbles

4) The price of 6 apples at the Quick Market is $2.7. The price of 7 of the same apples at Walmart is $3.64. Which place is the better buy?

5) The bakers at a Bakery can make 160 bagels in 4 hours. How many bagels can they bake in 11 hours? What is that rate per hour?

ISEE Lower-Level Subject Test Mathematics

✎ Answer each question and round your answer to the nearest whole number.

6) If a 24.6 ft tall flagpole casts a 190.95 ft long shadow, then how long is the shadow that a 4.7 ft tall woman casts?

7) A model igloo has a scale of 2 in: 7 ft. If the real igloo is 56 ft wide, then how wide is the model igloo?

8) If an 88 ft tall tree casts a 8 ft long shadow, then how tall is an adult giraffe that casts a 4 ft shadow?

9) Find the distance between San Joe and Mount Pleasant if they are 5 cm apart on a map with a scale of 1 cm: 8 km.

10) A telephone booth that is 54 ft tall casts a shadow that is 9 ft long. Find the height of a lawn ornament that casts a 7 ft shadow.

ISEE Lower-Level Subject Test Mathematics

Answers of Worksheets

Simplifying Ratios

1) 2: 8   7) 20: 1   13) 2: 3   19) 1: 25
2) 1: 3   8) 3: 2   14) 2: 5   20) 1: 10
3) 1: 7   9) 5: 4   15) 6: 1   21) 1: 3
4) 7: 6   10) 1: 2   16) 1: 2   22) 2: 9
5) 6: 7   11) 5: 7   17) 1: 11   23) 17: 20
6) 9: 7   12) 1: 9   18) 1: 4   24) 1: 10

Writing Ratios

1) $\frac{180 \text{ miles}}{6 \text{ gallons}}$, 30 miles per gallon

2) $\frac{99 \text{ dollars}}{11 \text{ books}}$, 9.00 dollars per book

3) $\frac{35 \text{ miles}}{3.5 \text{ gallons}}$, 10 miles per gallon

4) $\frac{7.5" \text{ of snow}}{1.5 \text{ hours}}$, 5 inches of snow per hour

5) $\frac{1}{10}$ 8) $\frac{2}{7}$ 11) $\frac{7}{24}$ 14) $\frac{1}{5}$

6) $\frac{2}{7}$ 9) $\frac{1}{7}$ 12) $\frac{2}{5}$ 15) $\frac{1}{3}$

7) $\frac{8}{15}$ 10) $\frac{2}{3}$ 13) $\frac{1}{10}$ 16) $\frac{1}{4}$

Create Proportion

1) 1: 5 = 4: 20   5) 9: 1 = 108: 12   9) 4: 5 = 36: 45
2) 9: 135 = 1: 15   6) 45: 18 = 5: 2   10) 128: 16 = 8: 1
3) 8: 5 = 24: 15   7) 24: 28 = 6: 7   11) 2.5: 5 = 10: 20
4) 7: 4 = 49: 28   8) 11: 3 = 55: 15   12) 15: 5 = 36: 12

Proportional Ratios

1) 10   7) 28   13) 240   19) 6
2) 2   8) 16.5   14) 100   20) 5
3) 9.6   9) 3.75   15) 14   21) 2
4) 27   10) 0.75   16) 4   22) 39
5) 100   11) 2   17) 8.4   23) 6
6) 12   12) 10.2   18) 2   24) 2

Similar Figures

1) 4   2) 2   3) 43

WWW.MathNotion.Com

ISEE Lower-Level Subject Test Mathematics

Word Problems

1) 300

2) The ratio for both classes is equal to 8 to 5.

3) Yes! Both ratios are 1 to 7

4) The price at the Quick Market is a better buy.

5) 440, the rate is 40 per hour.

6) 36.48 ft	8) 44 ft	10) 42 ft
7) 16 in	9) 40 km	

ISEE Lower-Level Subject Test Mathematics

Chapter 7 : Measurement

Topics that you'll learn in this chapter:

- ✓ Reference Measurement Units,
- ✓ Metric Length Units,
- ✓ Customary Length Units,
- ✓ Metric Capacity Units,
- ✓ Customary Capacity Units,
- ✓ Metric Weight and Mass Units,
- ✓ Customary Weight and Mass Units,
- ✓ Temperature Units,
- ✓ Time,
- ✓ Add Money Amounts,
- ✓ Subtract Money Amounts,
- ✓ Money: Word Problems,

ISEE Lower-Level Subject Test Mathematics

Reference Measurement Units

LENGTH

Customary

1 mile (mi) = 1,760 yards (yd)

1 yard (yd) = 3 feet (ft)

1 foot (ft) = 12 inches (in.)

Metric

1 kilometer (km) = 1,000 meters (m)

1 meter (m) = 100 centimeters (cm)

1 centimeter(cm) = 10 millimeters(mm)

VOLUME AND CAPACITY

Customary

1 gallon (gal) = 4 quarts (qt)

1 quart (qt) = 2 pints (pt.)

1 pint (pt.) = 2 cups (c)

1 cup (c) = 8 fluid ounces (Fl oz)

Metric

1 liter (L) = 1,000 milliliters (mL)

WEIGHT AND MASS

Customary

1 ton (T) = 2,000 pounds (lb.)

1 pound (lb.) = 16 ounces (oz)

Metric

1 kilogram (kg) = 1,000 grams (g)

1 gram (g) = 1,000 milligrams (mg)

Time

1 year = 12 months

1 year = 52 weeks

1 week = 7 days

1 day = 24 hours

1 hour = 60 minutes

1 minute = 60 seconds

ISEE Lower-Level Subject Test Mathematics

Metric Length Units

🖎 Convert to the units.

1) 300 mm = _____ cm

2) 8 m = _____ mm

3) 4.5 m = _____ cm

4) 7 km = _____ m

5) 9,400 mm = _____ m

6) 1,100 cm = _____ m

7) 2.8 m = _____ cm

8) 4,000 mm = _____ cm

9) 7,000 mm = _____ m

10) 2 km = _____ mm

11) 14.9 km = _____ m

12) 20 m = _____ cm

13) 5,000 m = _____ km

14) 7,600 m = _____ km

Customary Length Units

🖎 Convert to the units.

1) 8 ft = _____ in

2) 4 ft = _____ in

3) 6 yd = _____ ft

4) 10 yd = _____ ft

5) 3,520 yd = _____ mi

6) 60 in = _____ ft

7) 144 in = _____ yd

8) 0.5 mi = _____ yd

9) 15 yd = _____ in

10) 42 yd = _____ in

11) 99 ft = _____ yd

12) 1.5 mi = _____ yd

13) 84 in = _____ ft

14) 30 yd = _____ feet

ISEE Lower-Level Subject Test Mathematics

Metric Capacity Units

👉 Convert the following measurements.

1) 32.4 l = _____ ml

2) 7.1 l = _____ ml

3) 54 l = _____ ml

4) 92 l = _____ ml

5) 48 l = _____ ml

6) 13 l = _____ ml

7) 750 ml = _____ l

8) 2,400 ml = _____ l

9) 73,000 ml = _____ l

10) 8,000 ml = _____ l

11) 49,000 ml = _____ l

12) 5,500 ml = _____ l

Customary Capacity Units

👉 Convert the following measurements.

1) 51 gal = _____ qt.

2) 35 gal = _____ pt.

3) 68 gal = _____ c.

4) 20 pt. = _____ c

5) 12.5 qt = _____ pt.

6) 22.5 qt = _____ c

7) 51 pt. = _____ c

8) 48 c = _____ gal

9) 96 pt. = _____ gal

10) 136 qt = _____ gal

11) 15 c = _____ fl oz

12) 44 c = _____ qt

13) 240 c = _____ pt.

14) 148 qt = _____ gal

15) 160 pt. = _____ qt

16) 104 fl oz = _____ c.

WWW.MathNotion.Com

ISEE Lower-Level Subject Test Mathematics

Metric Weight and Mass Units

✎ Convert.

1) 60 kg = _____ g

2) 24 kg = _____ g

3) 610 kg = _____ g

4) 82 kg = _____ g

5) 95.8 kg = _____ g

6) 4.85 kg = _____ g

7) 1.5 kg = _____ g

8) 95,000 g = _____ kg

9) 241,000 g = _____ kg

10) 700,000 g = _____ kg

11) 2,500 g = _____ kg

12) 28,900 g = _____ kg

13) 970,000 g = _____ kg

14) 325,500 g = _____ kg

Customary Weight and Mass Units

✎ Convert.

1) 10,000 lb. = _____ T

2) 19,000 lb. = _____ T

3) 32,000 lb. = _____ T

4) 16,800 lb. = _____ T

5) 27 lb. = _____ oz

6) 25.4 lb. = _____ oz

7) 124 lb. = _____ oz

8) 4 T = _____ lb.

9) 7 T = _____ lb.

10) 11.2 T = _____ lb.

11) 12.8 T = _____ lb.

12) $\frac{6}{5}$ T = _____ oz

13) 9.125 T = _____ oz

14) $\frac{3}{4}$ T = _____ oz

WWW.MathNotion.Com

ISEE Lower-Level Subject Test Mathematics

Temperature Units

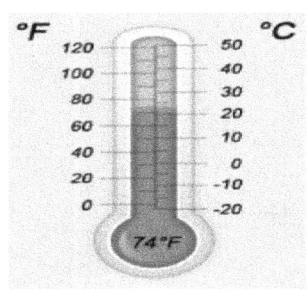

✎ Convert Fahrenheit into Celsius.

1) 21°F = ___ °C

2) 5.5°F = ___ °C

3) 71.6°F = ___ °C

4) 248°F = ___ °C

5) 75.2°F = ___ °C

6) 138.2°F = ___ °C

7) 167°F = ___ °C

8) 215.6°F = ___ °C

9) 104°F = ___ °C

10) 194°F = ___ °C

11) 203°F = ___ °C

12) 78.8°F = ___ °C

✎ Convert Celsius into Fahrenheit.

13) 2°C = ___ °F

14) 19°C = ___ °F

15) 48°C = ___ °F

16) 54°C = ___ °F

17) 81°C = ___ °F

18) 24°C = ___ °F

19) 88°C = ___ °F

20) 72°C = ___ °F

21) 80°C = ___ °F

22) 16°C = ___ °F

23) 10°C = ___ °F

24) 110°C = ___ °F

WWW.MathNotion.Com

ISEE Lower-Level Subject Test Mathematics

Time

Convert to the units.

1) 22 hr. = _____ min

2) 12.5 year = _____ week

3) 5.2 hr = _____ sec

4) 21 min = _____ sec

5) 1,800 min = _____ hr.

6) 730 day = _____ year

7) 1.5 year = _____ hr.

8) 42 day = _____ hr.

9) 2 day = _____ min

10) 660 min = _____ hr.

11) 10 year = _____ month

12) 2,124 sec = _____ min

13) 168 hr = _____ day

14) 19 weeks = _____ day

How much time has passed?

1) From 2:25 A.M. to 5:35 A.M.: ____ hours and ____ minutes.

2) From 3:30 A.M. to 7:55 A.M.: ____ hours and ____ minutes.

3) It's 6:30 P.M. What time was 2 hours ago? _____ O'clock

4) 4:30 A.M to 7:50 AM: _____ hours and _____ minutes.

5) 4:15 A.M to 8:35 AM: _____ hours and _____ minutes.

6) 5:10 A.M. to 7:35 AM. = _____ hour(s) and _____ minutes.

7) 10:55 A.M. to 3:25 PM. = _____ hour(s) and _____ minutes

8) 8:05 A.M. to 8:40 A.M. = _____ minutes

9) 6:02 A.M. to 6:49 A.M. = _____ minutes

WWW.MathNotion.Com

ISEE Lower-Level Subject Test Mathematics

Money Amounts

Add.

1) $128 + $328 $225 + $245 $150 + $186

2) $453 + $128 $440 + $541 $258 + $248

3) $645 + $112.5 $235.4 + $452.1 $125.99 + $148.32

4) $321.40 + $175.80 $458.10 + $752.65 $652.00 + $324.70

Subtract.

5) $725 − $334 $543 − $248 $349 − $122

6) $658.20 − $220.30 $752.10 − $452.15 $312.50 − $89.90

7) $315.90 − $220.10 $548.40 − $342.10 $968.40 − $324.50

8) Linda had $18.60. She bought some game tickets for $9.25. How much did she have left?

ISEE Lower-Level Subject Test Mathematics

Money: Word Problems

🖉 Solve.

1) How many boxes of envelopes can you buy with $40 if one box costs $8?

2) After paying $7.22 for a salad, Ella has $45.86. How much money did she have before buying the salad?

3) How many packages of diapers can you buy with $96 if one package costs $6?

4) Last week James ran 28.5 miles more than Michael. James ran 59 miles. How many miles did Michael run?

5) Last Friday Jacob had $14.68. Over the weekend he received some money for cleaning the attic. He now has $38.95. How much money did he receive?

6) After paying $3.15 for a sandwich, Amelia has $48.69. How much money did she have before buying the sandwich?

ISEE Lower-Level Subject Test Mathematics

Answers of Worksheets

Metric length
1) 30 cm
2) 8,000 mm
3) 450 cm
4) 7,000 m
5) 9.4 m
6) 11 m
7) 280 cm
8) 400 cm
9) 7 m
10) 2,000,000 mm
11) 14,900 m
12) 2,000 cm
13) 5 km
14) 7.6 km

Customary Length
1) 96
2) 48
3) 18
4) 30
5) 2
6) 5
7) 4
8) 880
9) 540
10) 1,512
11) 33
12) 2,640
13) 7
14) 90

Metric Capacity
1) 32,400 ml
2) 7,100 ml
3) 54,000 ml
4) 92,000 ml
5) 48,000 ml
6) 13,000 ml
7) 0.75ml
8) 2.4 ml
9) 73 ml
10) 8L
11) 49 L
12) 5.5 L

Customary Capacity
1) 204 qt
2) 280 pt.
3) 1,088 c
4) 40 c
5) 25 pt.
6) 90c
7) 102 c
8) 3 gal
9) 12 gal
10) 34 gal
11) 120 qt
12) 11qt
13) 120 pt.
14) 37 gal
15) 80 qt
16) 13 pt.

Metric Weight and Mass
1) 60,000 g
2) 24,000 g
3) 610,000 g
4) 82,000 g
5) 95,800g
6) 4,850 g
7) 1,500 g
8) 95 kg
9) 241 kg
10) 700 kg
11) 2.5 kg
12) 28.9 kg
13) 970 kg
14) 325.5 kg

Customary Weight and Mass
1) 5 T
2) 9.5 T
3) 16 T
4) 8.4 T
5) 432 oz
6) 406.4 oz

WWW.MathNotion.Com

ISEE Lower-Level Subject Test Mathematics

7) 1,984 oz
8) 8,000 lb.
9) 14,000 lb.
10) 22,400 lb.
11) 25,600 lb.
12) 38,400 oz
13) 292,000 oz
14) 24,000 oz

Temperature

1) −6.11°C
2) −14.72°C
3) 22°C
4) 120°C
5) 24°C
6) 59°C
7) 75°C
8) 102°C
9) 40°C
10) 90°C
11) 95°C
12) 26°C
13) 35.6°F
14) 66.2°F
15) 118.4°F
16) 129.2°F
17) 177.8°F
18) 75.2°F
19) 190.4°F
20) 161.6°F
21) 176°F
22) 60.8°F
23) 50°F
24) 230°F

Time - Convert

1) 1,320 min
2) 650 weeks
3) 18,720 sec
4) 1,260 sec
5) 30 hr
6) 2 year
7) 13,140 hr
8) 1,008 hr
9) 2,880 min
10) 11 hr
11) 120 months
12) 35.4 min
13) 7 days
14) 133 days

Time - Gap

1) 3:10
2) 4:25
3) 4:30 P.M.
4) 3:20
5) 4:20
6) 2:25
7) 4:30
8) 35 minutes
9) 47 minutes

Add Money

1) 456, 470, 336
2) 581, 981, 506
3) 757.5, 687.5, 274.31
4) 497.2, 1210.75, 976.7

Subtract Money

5) 391, 295, 227
6) 437.9, 299.95, 222.6
7) 95.8, 206.3, 643.9
8) $9.35

Money: word problem

1) 5
2) $53.08
3) 16
4) 30.5
5) 24.27
6) 51.84

Chapter 8 : Algebraic Thinking

Topics that you'll learn in this chapter:

- ✓ Finding Rules,
- ✓ Algebraic Word Problems,
- ✓ Evaluating Expressions,
- ✓ Variables and Expressions,

Finding Rules

Complete the output.

1- **Rule:** the output is $x - 10.5$

Input	x	15	18	27	32.25	48.5
Output	y					

17) **Rule:** the output is $x \times 5\frac{1}{3}$

Input	x	3	9	15	21	33
Output	y					

2- **Rule:** the output is $x \div 9$

Input	x	513	387	342	198	126
Output	y					

Find a rule to write an expression.

3- **Rule:** _____

Input	x	4	14	19	24
Output	y	10	35	47.5	60

4- **Rule:** _____

Input	x	5	13	19.6	34.5
Output	y	14.4	22.4	29	43.9

5- **Rule:** _____

Input	x	72	96	132	230.4
Output	y	9	12	16.5	28.8

ISEE Lower-Level Subject Test Mathematics

Algebraic Word Problems

Circle the number sentence that fits the problem. Then solve for x.

1) Mary had $42. Then she earned more money (x). Now she has $86.

   $42 + x = $86   OR $42 + $86 = x

   x = ____

2) Lisa had $35. Then she earned more money (x). Now she has $78.

   $35 + x = $78   OR $35 + $78 = x

   x = ____

3) Matthew had $37. Then he earned more money (x). Now he has $98.

   $37 + x = $98   OR $37 + $98 = x

   x = ____

4) Charlotte gave 19 of the cookies he had baked to a friend and now he has 45 cookies left.   45 − 19 = x   OR   x − 19 = 45

   x = ____

5) Mia gave 32 of the cookies she had baked to a friend and now she has 55 cookies left.   55 − 32 = x   OR   x − 32 = 55

   x = ____

6) Lucas gave 41 of the cookies he had baked to a friend and now he has 49 cookies left. . 49 − 41 = x   OR   x − 41 = 49

   x = ____

WWW.MathNotion.Com

Evaluate Expressions

✎Simplify each algebraic expression.

1) $11 - x$, $x = 3$

2) $x + 14$, $x = 4$

3) $8 - 3x$, $x = 1$

4) $3x + \frac{1}{3}$, $x = \frac{1}{2}$

5) $2x + 18$, $x = 1.5$

6) $7 - 3x$, $x = 1.2$

7) $12 + 2x - 15$, $x = 3.5$

8) $25 - 5x$, $x = 2.2$

9) $\frac{44}{x} - 58$, $x = 0.4$

10) $\frac{x}{3} - 15 + x$, $x = 12.6$

11) $\frac{x}{7} + 9.5$, $x = 28.7$

12) $\frac{33}{x} - 5.2 + 2.1x$, $x = 3$

13) $2x - \frac{45}{x} - 12$, $x = 9$

14) $\frac{x}{17} - 1.8$, $x = 34$

15) $2(12.5x + 8)$, $x = 2$

16) $16x + 13x - 27 + 9$,

   $x = 1.5$

17) $8.7 - \frac{16}{x} + 3x$,

   $x = 4$

18) $5(3a - 2a)$,

   $a = 5.5$

19) $14 - 2x + 16 - x$,

   $x = 3.5$

20) $6x - 3 - x$,

   $x = 1.6$

21) $18 - 2(2x + x)$, $x = 0.2$

ISEE Lower-Level Subject Test Mathematics

Variables and Expressions

✎ Write a verbal expression for each algebraic expression.

1) $3a - 7b$

2) $8.2c^2 + 5d$

3) $x - 19.5$

4) $\dfrac{90}{6}$

5) $x^2 + y^3$

6) $4x + 9$

7) $x^2 - 6y + 15$

8) $x^3 + 7y^2 - 6$

9) $\dfrac{1}{5}x + \dfrac{1}{4}y - 11$

10) $\dfrac{1}{7}(x + 13) - 18.6y$

✎ Write an algebraic expression for each verbal expression.

11) 14 less than h

12) The product of 19 and a

13) The 23.2 divided by K

14) The product of 7 and the third power of x

15) 14 more than h to the sixth power

16) 30 more than triple d

17) One eighth the square of b

18) The difference of 42.5 and 3 times a number

19) 73 more than the cube of a number

20) one-quarters the cube of a number

WWW.MathNotion.Com

ISEE Lower-Level Subject Test Mathematics

Answers of Worksheets

Finding Rules

1)

Input	x	15	18	27	32.25	48.5
Output	y	4.5	7.5	16.5	21.75	38

2)

Input	x	3	9	15	21	33
Output	y	16	48	80	112	176

3)

Input	x	513	387	342	198	126
Output	y	57	43	38	22	14

4) $y = 2.5x$ 5) $y = x + 9.4$ 6) $y = x \div 8$

Algebraic Word Problems

1) $\$42 + x = \$86$; $x = 44$ 4) $x - 19 = 45$; $x = 64$

2) $\$35 + x = \$78$; $x = 43$ 5) $x - 32 = 55$; $x = 87$

3) $\$37 + x = \$98$; $x = 61$ 6) $x - 41 = 49$; $x = 90$

Evaluating Expressions

1) 8
2) 18
3) 5
4) $\frac{11}{6}$
5) 21
6) 3.4
7) 4
8) 14
9) 52
10) 1.8
11) 13.6
12) 12.1
13) 1
14) 0.2
15) 66
16) 25.5
17) 16.7
18) 27.5
19) 19.5
20) 5
21) 16.8

Variables and Expressions

1) 3 times a minus 7 times b.

2) 8.2 times c squared plus 5 times d.

3) a number minus 19.5.

4) the quotient of 90 and 6.

5) x squared plus y cubed.

6) the product of 4 and x plus 9.

WWW.MathNotion.Com

ISEE Lower-Level Subject Test Mathematics

7) x squared plus the product of 6 and y plus 15.

8) x cubed plus the product of 7 and y squared minus the product of 6 and y.

9) the sum of one–fifth of x and one–quarters of y, minus 11.

10) one–seventh of the sum of x and 13 minus the product of 18.6 and y.

11) $14 < h$

12) $19a$

13) $\frac{23.2}{K}$

14) $7x^3$

15) $14 > h^6$

16) $3d < 30$

17) $\frac{1}{8}b^2$

18) $42.5 - 3a$

19) $73 > a^3$

20) $\frac{1}{4}x^3$

ISEE Lower-Level Subject Test Mathematics

Chapter 9 : Geometric

Topics that you'll learn in this chapter:

- ✓ Identifying Angles,
- ✓ Estimate and Measure Angles,
- ✓ Polygon Names,
- ✓ Classify Triangles,
- ✓ Parallel Sides in Quadrilaterals,
- ✓ Identify Parallelograms,
- ✓ Identify Trapezoids,
- ✓ Identify Rectangles,
- ✓ Perimeter and Area of Squares,
- ✓ Perimeter and Area of rectangles,
- ✓ Area and Perimeter: Word Problems,
- ✓ Circumference, Diameter and Radius,
- ✓ Volume of Cubes and Rectangle Prisms,

ISEE Lower-Level Subject Test Mathematics

Identifying Angles

✎ Write the name of the angles (Acute, Right, Obtuse, and Straight).

1)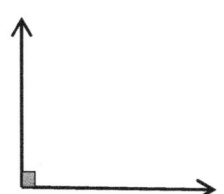

2)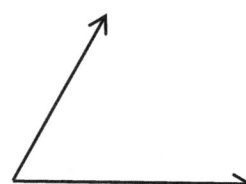

3)

4)

5)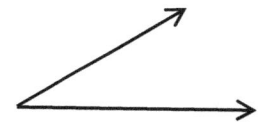

6)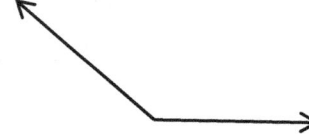

7)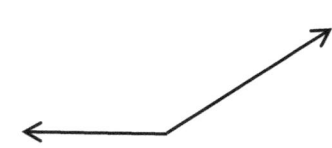

8)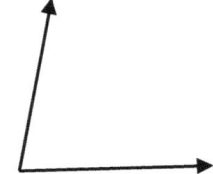

ISEE Lower-Level Subject Test Mathematics

Estimate Angle Measurements

✏️ Estimate the approximate measurement of each angle in degrees.

1)

2)

3)
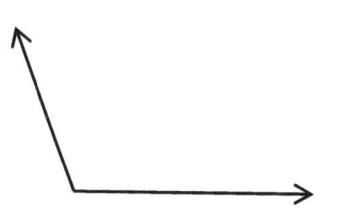

4)
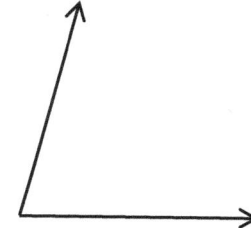

5)
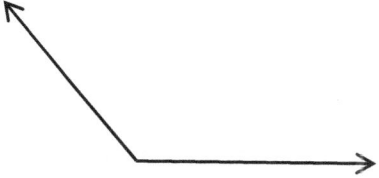

6)
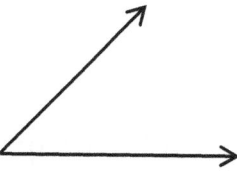

7)
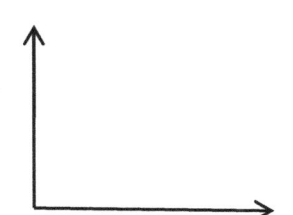

8)
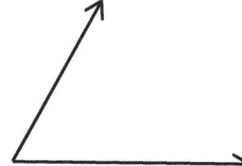

ISEE Lower-Level Subject Test Mathematics

Measure Angles with a Protractor

✎ Use protractor to measure the angles below.

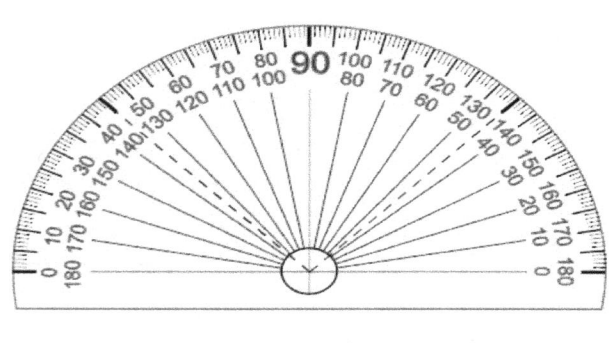

1)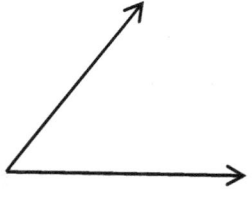

2)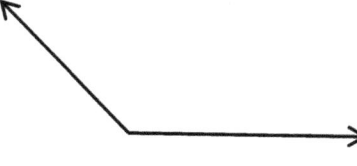

3)

4)

✎ Use a protractor to draw angles for each measurement given.

1) 140°

2) 100°

3) 110°

4) 120°

5) 55°

ISEE Lower-Level Subject Test Mathematics

Polygon Names

✎ Write name of polygons.

1)

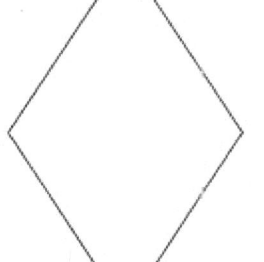

2)

3)

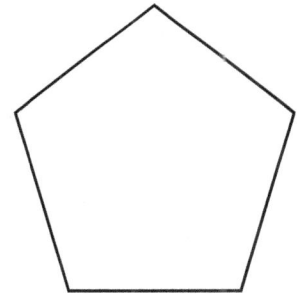

4)

5)

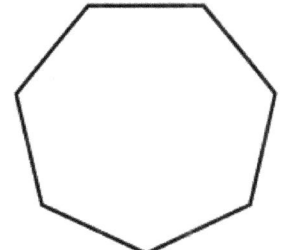

6)

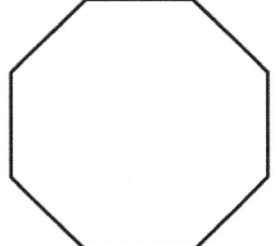

WWW.MathNotion.Com   93

Classify Triangles

✏ Classify the triangles by their sides and angles.

1)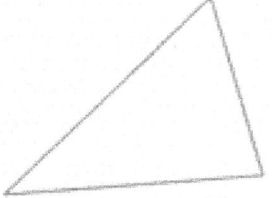

2)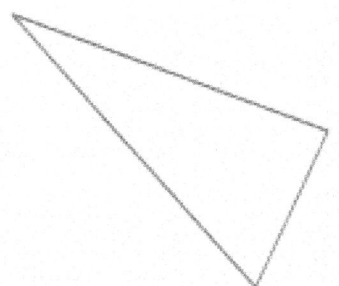

3)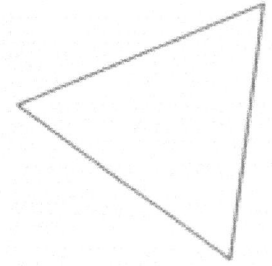

4)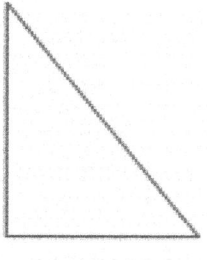

5)

6)

ISEE Lower-Level Subject Test Mathematics

Parallel Sides in Quadrilaterals

✎ Write name of quadrilaterals.

1)

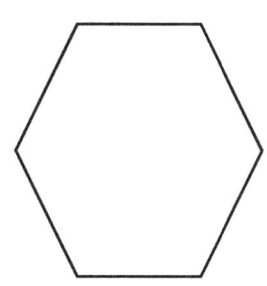

2)

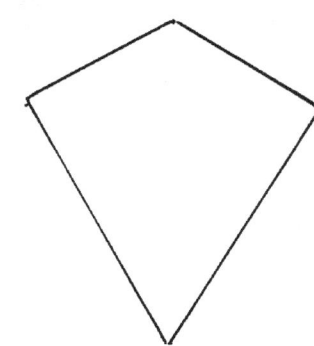

3)

4)

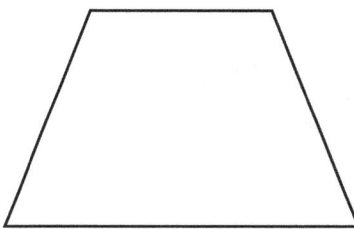

5)

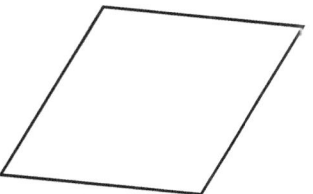

6)

> **ISEE Lower-Level Subject Test Mathematics**

Identify Rectangles

✏️ Solve.

1) A rectangle has _____ sides and _____ angles.

2) Draw a rectangle that is 5.5 centimeters long and 2.5 centimeters wide. What is the perimeter?

3) Draw a rectangle 3.5 cm long and 1.5 cm wide.

4) Draw a rectangle whose length is 4.25 cm and whose width is 2.45 cm. What is the perimeter of the rectangle?

5) What is the perimeter of the rectangle?

7.2

5.8

ISEE Lower-Level Subject Test Mathematics

Perimeter: Find the Missing Side Lengths

✎ Find the missing side of each shape.

1) perimeter = 57.2

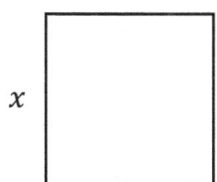

2) perimeter = 21.2

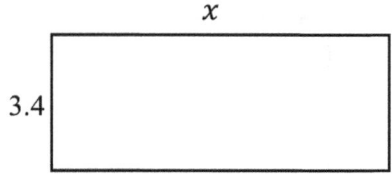

3) perimeter = 27.5

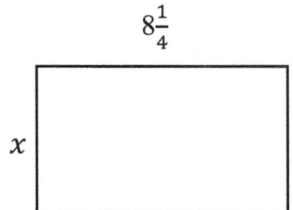

4) perimeter = 35.2

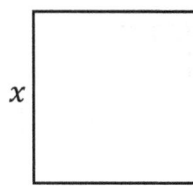

5) perimeter = 75.6

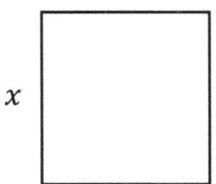

6) perimeter = 30.8

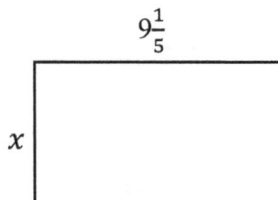

7) perimeter = 36.25

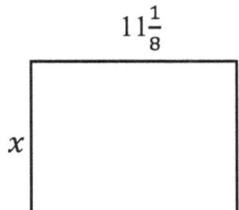

8) perimeter = 46.8

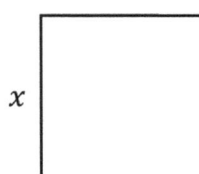

WWW.MathNotion.Com

ISEE Lower-Level Subject Test Mathematics

Perimeter and Area of Squares

✏ Find perimeter and area of squares.

1) A: _____, P: _____

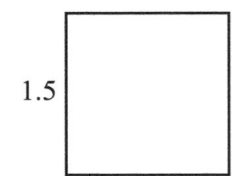
1.5

2) A: _____, P: _____

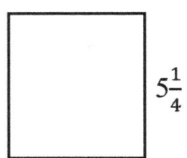
$5\frac{1}{4}$

3) A: _____, P: _____

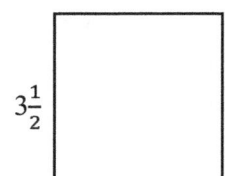
$3\frac{1}{2}$

4) A: _____, P: _____

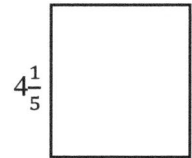
$4\frac{1}{5}$

5) A: _____, P: _____

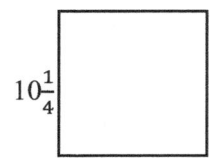
$10\frac{1}{4}$

6) A: _____, P: _____

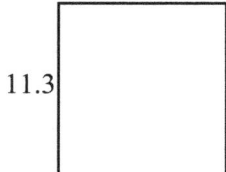
11.3

7) A: _____, P: _____

12.4

8) A: _____, P: _____

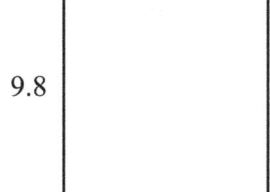
9.8

WWW.MathNotion.Com

ISEE Lower-Level Subject Test Mathematics

Perimeter and Area of rectangles

🖎 Find perimeter and area of rectangles.

1) A: _____, P: _____

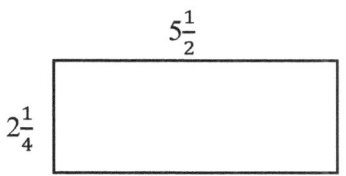

2) A: _____, P: _____

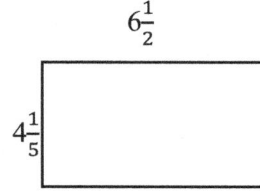

3) A: _____, P: _____

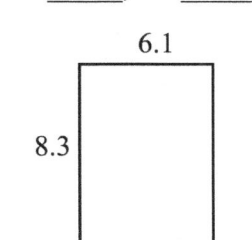

4) A: _____, P: _____

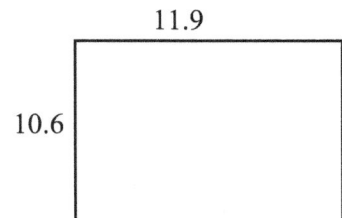

5) A: _____, P: _____

6) A: _____, P: _____

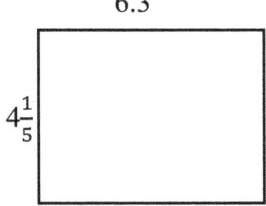

7) A: _____, P: _____

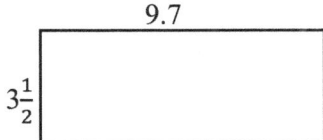

8) A: _____, P: _____

www.MathNotion.Com

ISEE Lower-Level Subject Test Mathematics

Find the Area or Missing Side Length of a Rectangle

👉 Find area or missing side length of rectangles.

1) Area =?

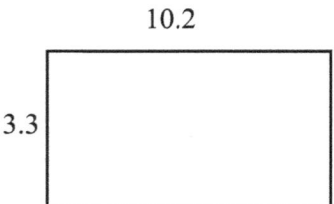

2) Area = 42.12, x=?

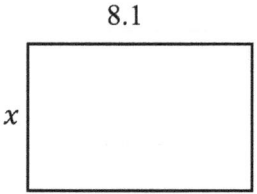

3) Area = 29.52, x=?

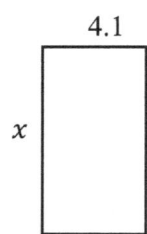

4) Area =?

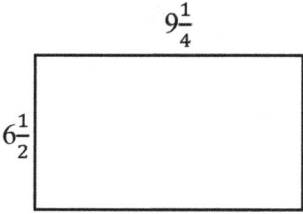

5) Area =?

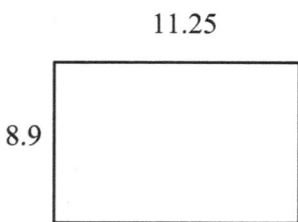

6) Area = 662.34 x=?

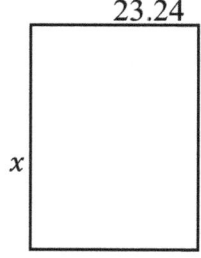

7) Area = 216.24, x=?

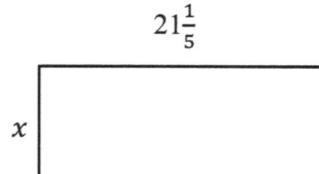

8) Area 336.42, x=?

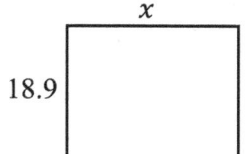

WWW.MathNotion.Com

ISEE Lower-Level Subject Test Mathematics

Area and Perimeter: Word Problems

✎ Solve.

1) The area of a rectangle is 90.86 square meters. The width is 7.7 meters. What is the length of the rectangle?

2) A square has an area of 6.25 square feet. What is the perimeter of the square?

3) Ava built a rectangular vegetable garden that is 3.2 feet long and has an area of 21.12 square feet. What is the perimeter of Ava's vegetable garden?

4) A square has a perimeter of 12.8 millimeters. What is the area of the square?

5) The perimeter of David's square backyard is 0.96 meters. What is the area of David's backyard?

6) The area of a rectangle is 37.63 square inches. The length is 7.1 inches. What is the perimeter of the rectangle?

ISEE Lower-Level Subject Test Mathematics

Circumference, Diameter, and Radius

🖎 Find the diameter and circumference of circles.

1)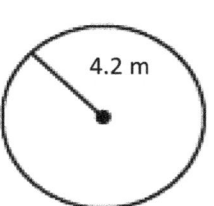

2)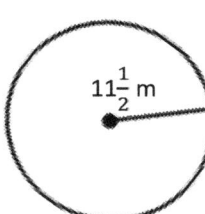

3)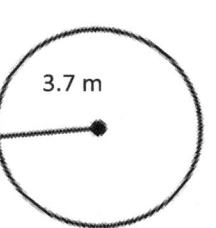

4)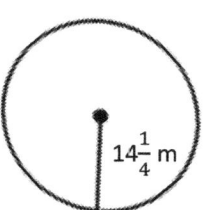

🖎 Find the radius.

5)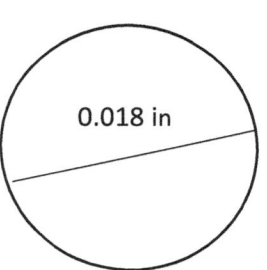

6)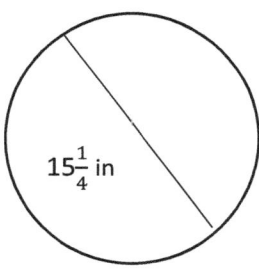

7) Diameter = $16\frac{1}{5}$ ft

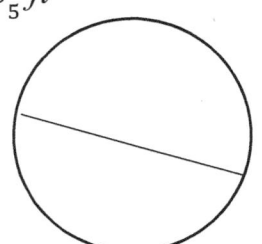

8) Diameter = 23.82 m

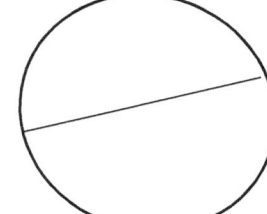

WWW.MathNotion.Com

ISEE Lower-Level Subject Test Mathematics

Volume of Cubes and Rectangle Prisms

✏️ Find the volume of each of the rectangular prisms.

1)

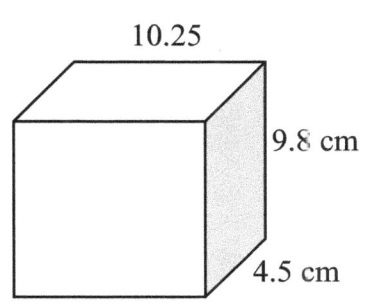

2)

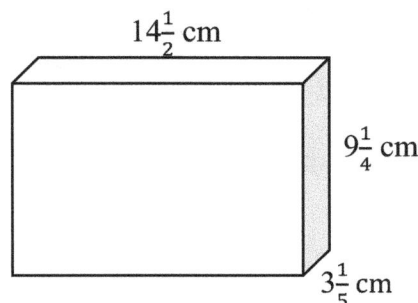

3)

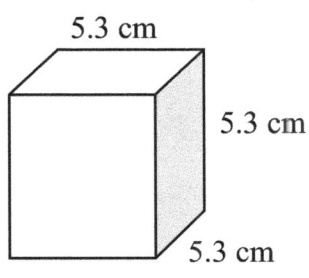

4)

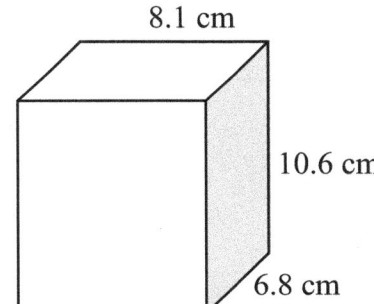

5)

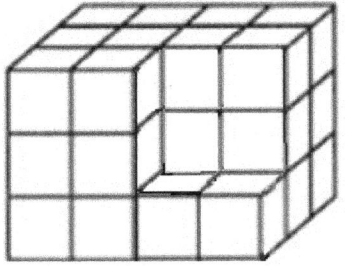

6)

WWW.MathNotion.Com

ISEE Lower-Level Subject Test Mathematics

Answers of Worksheets

Identifying Angles

1) Right   3) Obtuse   5) Acute   7) Obtuse

2) Acute   4) Straight   6) Obtuse   8) Acute

Estimate Angle Measurements

1) 160°   3) 110°   5) 130°   7) 90°

2) 180°   4) 75°   6) 45°   8) 60°

Measure Angles with a Protractor

1) 50°   2) 135°   3) 20°   4) 170°

Draw Angles

1)  2) 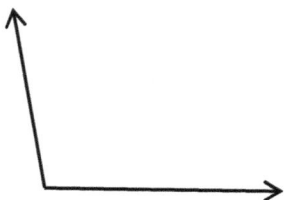 3)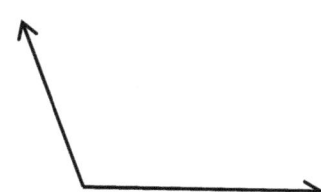

4) 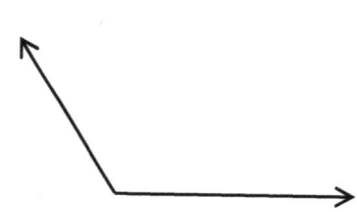   5)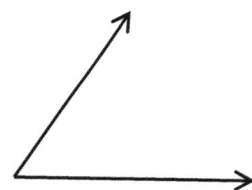

Polygon Names

1) Diamond   3) Pentagon   5) Heptagon

2) Parallelogram   4) Trapezius   6) Octagon

Classify Triangles

1) Scalene, acute   4) Scalene, right

2) Isosceles, acute   5) Isosceles, right

3) Equilateral, acute   6) Scalene, obtuse

WWW.MathNotion.Com

ISEE Lower-Level Subject Test Mathematics

Parallel Sides in Quadrilaterals

1) Hexagon
2) Kike
3) Parallelogram
4) Trapezoid
5) Rhombus
6) Rectangle

Identify Rectangles

1) 4 - 4
2) 16
3) Draw the rectangle.
4) 13.4
5) 26

Perimeter: Find the Missing Side Lengths

1) 14.3
2) 7.2
3) 5.5
4) 8.8
5) 18.9
6) 6.2
7) 7
8) 11.7

Perimeter and Area of Squares

1) A: 2.25, P: 6
2) A: 27.56, P: 21
3) A: 12.25, P: 14
4) A: 17.64, P: 16.8
5) A: 105.063 P: 41
6) A: 127.69, P: 45.2
7) A: 153.76, P: 49.6
8) A: 96.04, P: 39.2

Perimeter and Area of rectangles

1) A: 12.375, P: 15.5
2) A: 27.3, P: 21.4
3) A: 50.63, P: 28.8
4) A: 126.14, P: 45
5) A: 39.9, P: 25.7
6) A: 26.46, P: 21
7) A: 33.95, P: 26.4
8) A: 66.12, P: 34.4

Find the Area or Missing Side Length of a Rectangle

1) 33.66
2) 5.2
3) 7.2
4) 60.125
5) 100.125
6) 28.5
7) 10.2
8) 17.8

Area and Perimeter: Word Problems

1) 11.8
2) 10
3) 19.6
4) 10.24
5) 0.0576
6) 24.8

Circumference, Diameter, and Radius

1) diameter: 8.4 circumferences: 8.4π or 26.376
2) diameter: 23 circumferences: 23π or 72.22
3) diameter: 7.4 circumferences: 7.4π or 23.24
4) diameter: 28.5 circumferences: 28.5π or 89.49
5) radius: 0.009 in
6) radius: 7.625 in
7) radius: 8.1 ft
8) radius: 11.91 m

Volume of Cubes and Rectangle Prisms

1) 452.025 cm^3
2) 429.2 cm^3
3) 148.877 c m^3
4) 583.848 cm^3
5) 32
6) 40

Chapter 10 : Three-Dimensional Figures

Topics that you'll learn in this chapter:

- ✓ Identify Three–Dimensional Figures,
- ✓ Count Vertices, Edges, and Faces,
- ✓ Identify Faces of Three–Dimensional Figures,

ISEE Lower-Level Subject Test Mathematics

Identify Three-Dimensional Figures

✎ Write the name of each shape.

1)

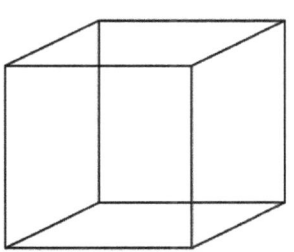

2)

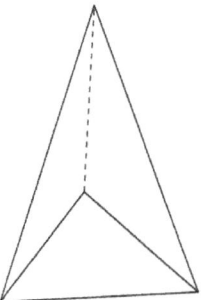

3)

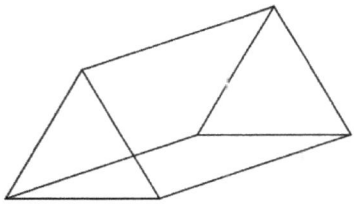

4)

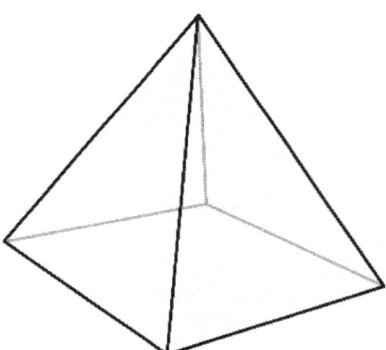

5)

6)

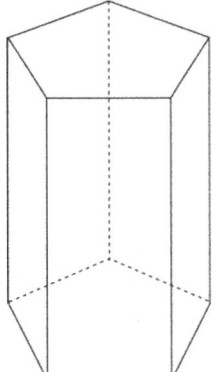

WWW.MathNotion.Com

ISEE Lower-Level Subject Test Mathematics

Count Vertices, Edges, and Faces

	Shape	Number of edges	Number of faces	Number of vertices
1)		___	___	___
2)		___	___	___
3)		___	___	___
4)		___	___	___
5)		___	___	___
6)		___	___	___

Identify Faces of Three–Dimensional Figures

✍ Write the number of faces.

1)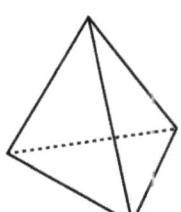

2)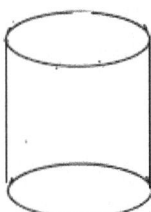

3)

4)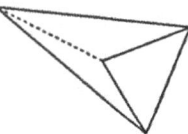

5)

6)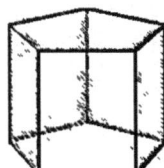

7)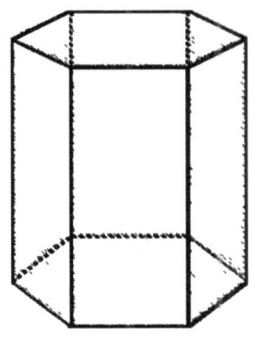

8)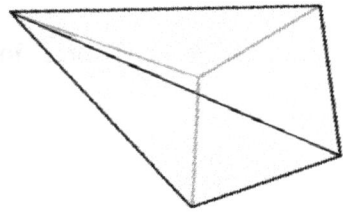

ISEE Lower-Level Subject Test Mathematics

Answers of Worksheets

Identify Three–Dimensional Figures

1) Cube
2) Triangular pyramid
3) Triangular prism
4) Square pyramid
5) Rectangular prism
6) Pentagonal prism
7) Hexagonal prism

Count Vertices, Edges, and Faces

Shape	Number of edges	Number of faces	Number of vertices
1)	6	4	4
2)	8	5	5
3)	12	6	8
4)	15	7	10
5)	12	6	8
6)	18	8	12

Identify Faces of Three–Dimensional Figures

1) 6
2) 2
3) 5
4) 4
5) 6
6) 7
7) 8
8) 5

ISEE Lower-Level Subject Test Mathematics

Chapter 11 : Symmetry and Transformations

Topics that you'll learn in this chapter:

- ✓ Line Segments,
- ✓ Identify Lines of Symmetry,
- ✓ Count Lines of Symmetry,
- ✓ Parallel, Perpendicular and Intersecting Lines,

ISEE Lower-Level Subject Test Mathematics

Line Segments

✏️ Write each as a line, ray, or line segment.

1)

2)

3)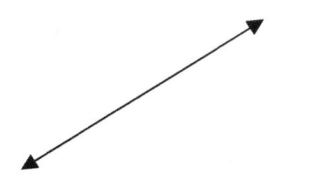

4)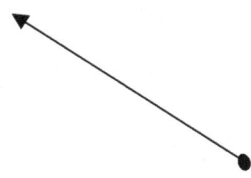

5)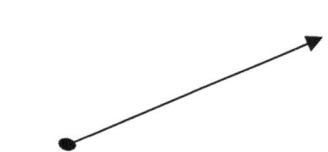

6)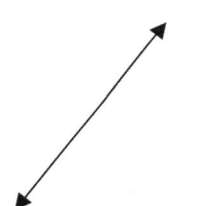

7)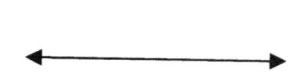

8)

WWW.MathNotion.Com

ISEE Lower-Level Subject Test Mathematics

Identify Lines of Symmetry

✍ Tell whether the line on each shape a line of symmetry is.

1)

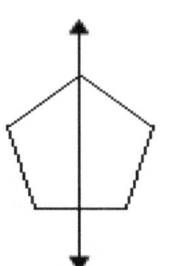

2)

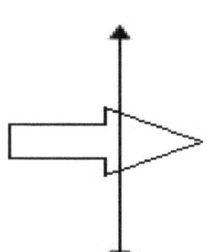

3)

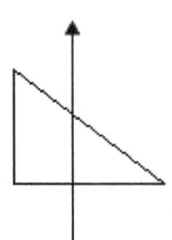

4)

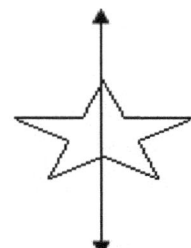

5)

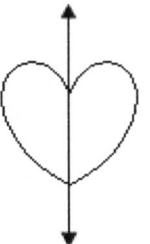

6)

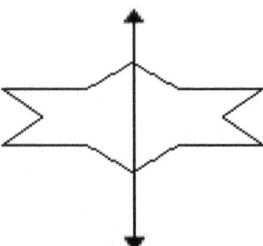

7)

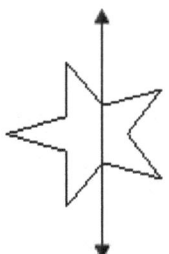

8)

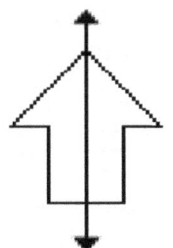

WWW.MathNotion.Com

ISEE Lower-Level Subject Test Mathematics

Count Lines of Symmetry

✍ Draw lines of symmetry on each shape. Count and write the lines of symmetry you see.

1)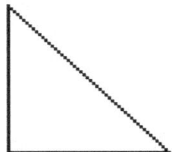

2)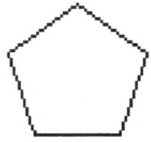

3)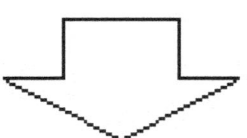

4)

5)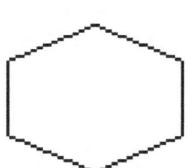

6)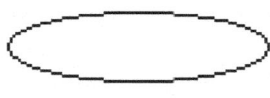

7)

8)

ISEE Lower-Level Subject Test Mathematics

Parallel, Perpendicular and Intersecting Lines

🖊️ State whether the given pair of lines are parallel, perpendicular, or intersecting.

1)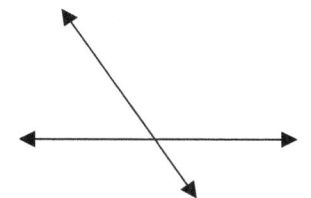

2)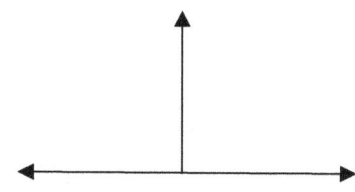

3)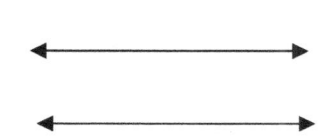

4)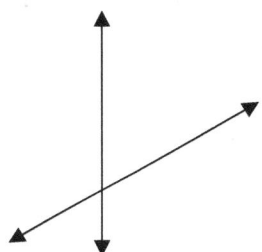

5)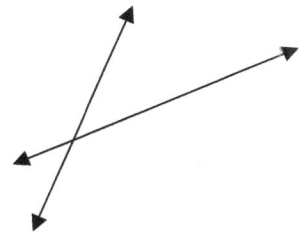

6)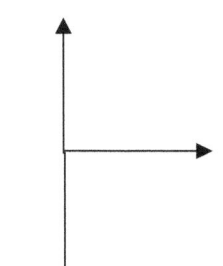

7)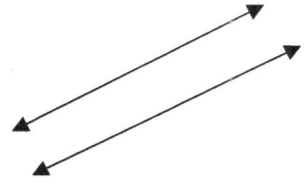

8)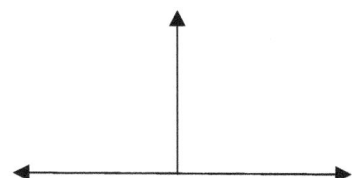

ISEE Lower-Level Subject Test Mathematics

Answers of Worksheets

Line Segments

1) Ray
2) Line segment
3) Line
4) Ray
5) Ray
6) Line
7) Line
8) Line segment

Identify lines of symmetry

1) yes
2) no
3) no
4) yes
5) yes
6) yes
7) no
8) yes

Count lines of symmetry

1) 2) 3) 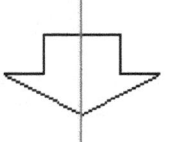 4)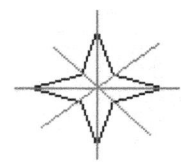

5) 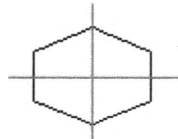 6) 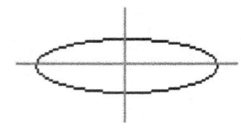 7) 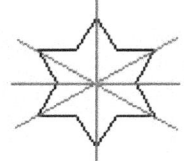 8)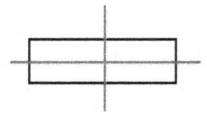

Parallel, Perpendicular and Intersecting Lines

1) Intersection
2) Perpendicular
3) Parallel
4) Intersection
5) Intersection
6) Perpendicular
7) Parallel
8) Perpendicular

Chapter 12 : Data Graphs, and Statistics

Topics that you'll learn in this chapter:

- ✓ Mean, Median, Mode, and Range,
- ✓ Graph Points on a Coordinate Plane,
- ✓ Bar Graph,
- ✓ Tally and Pictographs,
- ✓ Dot Plots
- ✓ Line Graphs,
- ✓ Stem-And-Leaf Plot,
- ✓ Scatter Plots,
- ✓ Probability Problems,

ISEE Lower-Level Subject Test Mathematics

Mean and Median

☛ Find Mean and Median of the Given Data.

1) 13, 22, 12, 6, 14

2) 7, 18, 11, 15, 2, 23

3) 33, 25, 14, 7, 14

4) 6, 7, 5, 1, 2, 5

5) 11, 4, 15, 7, 8, 17, 22

6) 9, 2, 5, 2, 18, 14, 45

7) 19, 14, 20, 12, 34, 14, 17, 31

8) 38, 29, 5, 3, 14, 9, 32

9) 28, 27, 31, 36, 32, 53, 41

10) 12, 17, 6, 17, 2, 17, 9, 12

11) 35, 11, 25, 54, 42, 41

12) 42, 48, 50, 48, 27, 67

13) 75, 72, 30, 46, 38, 29

14) 89, 73, 67, 46, 54, 84, 34

15) 78, 21, 100, 85, 54, 60

16) 41, 65, 9, 88, 17, 38, 14

☛ Solve.

17) In a javelin throw competition, five athletics score 54, 72, 59, 67 and 85 meters. What are their Mean and Median? _____

18) Eva went to shop and bought 18 apples, 7 peaches, 6 bananas, 3 pineapple and 9 melons. What are the Mean and Median of her purchase?

_____

WWW.MathNotion.Com

ISEE Lower-Level Subject Test Mathematics

Mode and Range

✏️ Find Mode and Rage of the Given Data.

1) 12, 4, 8, 2, 9, 4
   Mode: _____ Range: _____

2) 8, 8, 11, 8, 18, 2, 5, 41
   Mode: _____ Range: _____

3) 3, 3, 2, 23, 4, 12, 3, 9, 3
   Mode: _____ Range: _____

4) 12, 39, 5, 25, 4, 4, 27, 8
   Mode: _____ Range: _____

5) 5, 5, 8, 5, 14, 3, 12
   Mode: _____ Range: _____

6) 2, 5, 19, 15, 12, 11, 7, 8, 7, 7
   Mode: _____ Range: _____

7) 9, 4, 0, 12, 19, 21, 9, 7, 9, 3
   Mode: _____ Range: _____

8) 11, 4, 3, 11, 5, 8, 44, 11, 7
   Mode: _____ Range: _____

9) 3, 3, 6, 9, 3, 3, 9, 15, 14, 20
   Mode: _____ Range: _____

10) 15, 14, 14, 16, 20, 7, 1, 30
    Mode: _____ Range: _____

11) 13, 3, 27, 4, 4, 16, 18, 4
    Mode: _____ Range: _____

12) 7, 22, 35, 11, 7, 24, 6, 13
    Mode: _____ Range: _____

✏️ Solve.

13) A stationery sold 14 pencils, 28 red pens, 51 blue pens, 14 notebooks, 27 erasers, 43 rulers and 40 color pencils. What are the Mode and Range for the stationery sells?

Mode: _____   Range: _____

14) In an English test, eight students score 14, 20, 22, 14, 17, 28, 36 and 24. What are their Mode and Range? _____

WWW.MathNotion.Com

ISEE Lower-Level Subject Test Mathematics

Graph Points on a Coordinate Plane

🖎 Plot each point on the coordinate grid.

1) A (4, 6) 3) C (1, 5) 5) E (4, 8)
2) B (3, 2) 4) D (5, 7) 6) F (9, 2)

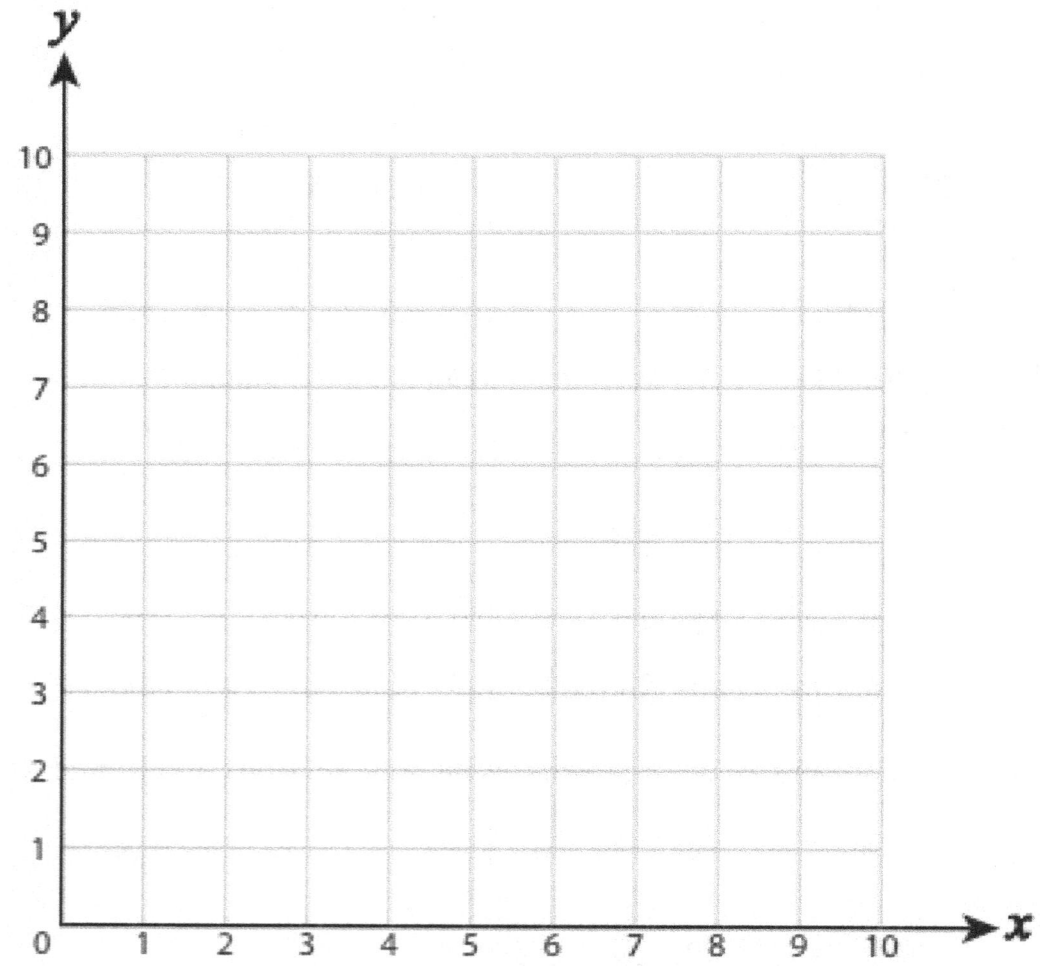

WWW.MathNotion.Com

Bar Graph

🖉 Graph the given information as a bar graph.

Day	Hot dogs sold
Monday	50
Tuesday	80
Wednesday	10
Thursday	30
Friday	70

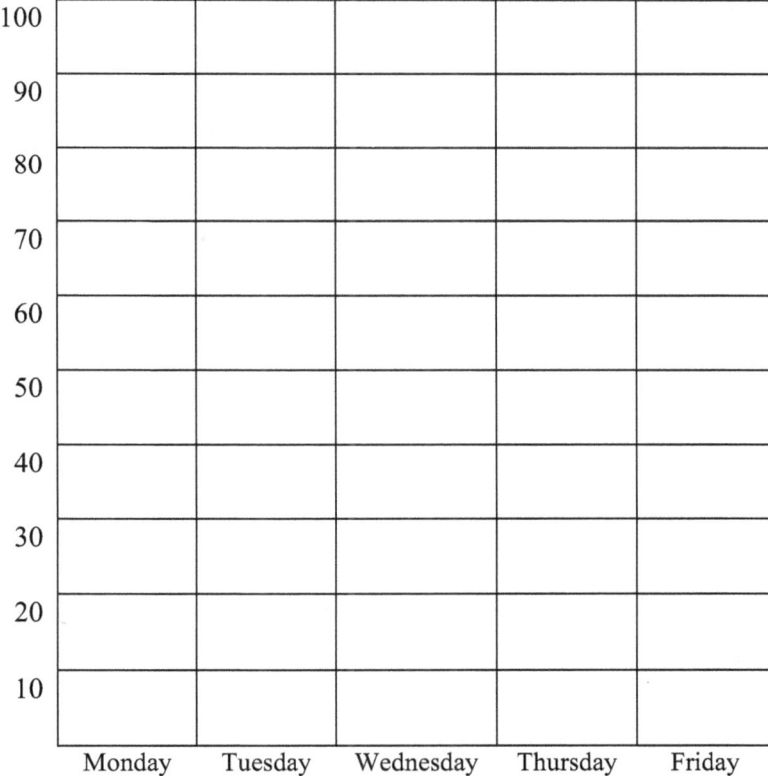

ISEE Lower-Level Subject Test Mathematics

Tally and Pictographs

 Using the key, draw the pictograph to show the information.

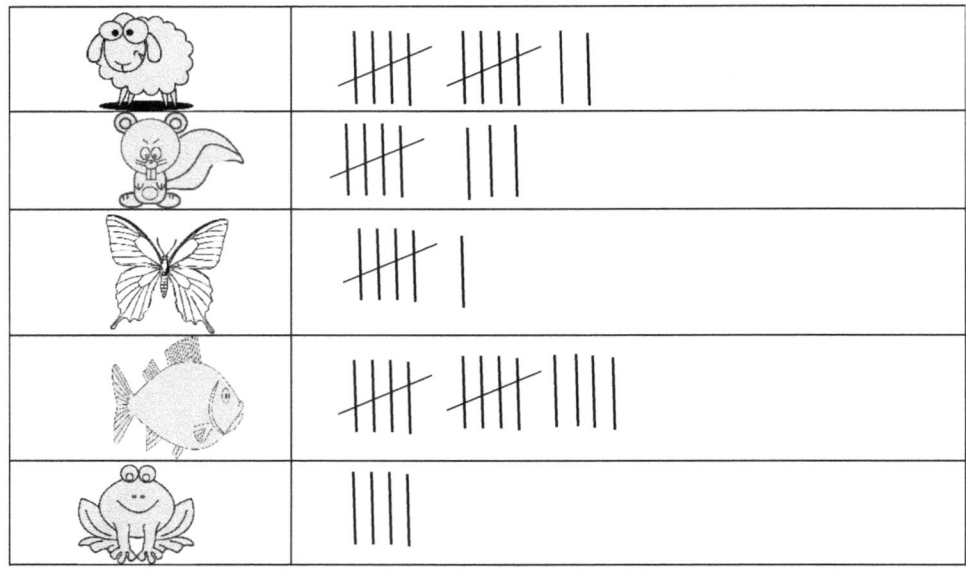

Key: 🙂 = 2 animals

Dot plots

The ages of students in a Math class are given below.

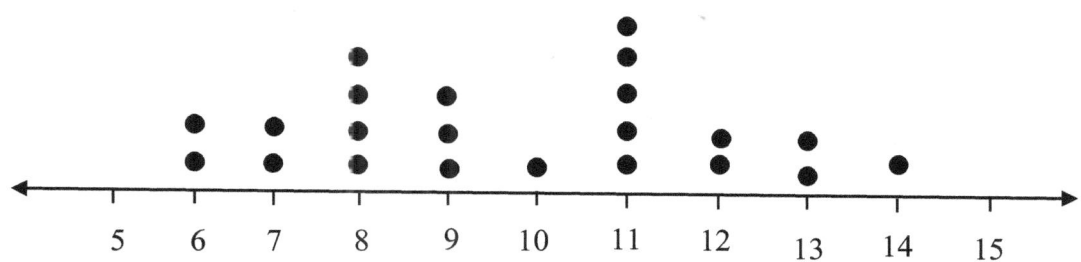

1) What is the total number of students in math class?

2) How many students are at least 12 years old?

3) Which age(s) has the most students?

4) Which age(s) has the fewest student?

5) Determine the median of the data.

6) Determine the range of the data.

7) Determine the mode of the data.

ISEE Lower-Level Subject Test Mathematics

Line Graphs

David work as a salesman in a store. He records the number of shoes sold in five days on a line graph. Use the graph to answer the question

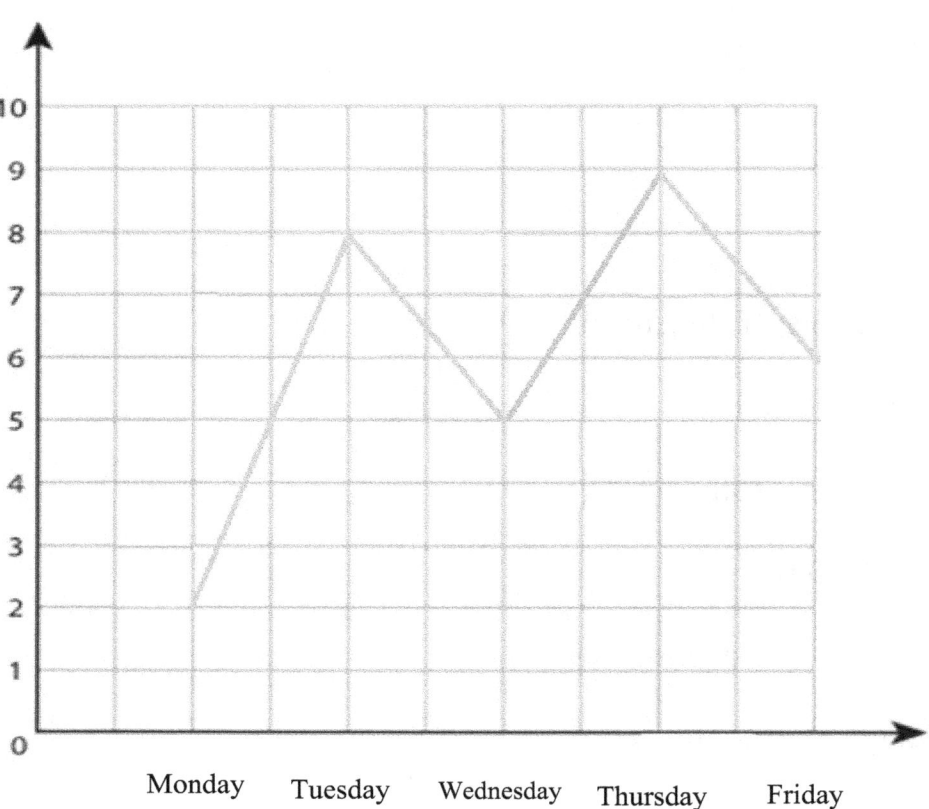

1) How many shoes were sold on Friday?

2) Which day had the minimum sales of shoes?

3) Which day had the maximum number of shoes sold?

4) How many shoes were sold in 5 days?

WWW.MathNotion.Com

ISEE Lower-Level Subject Test Mathematics

Stem–And–Leaf Plot

 Make stem ad leaf plots for the given data.

1) 42, 47, 14, 19, 42, 69, 65, 49, 42, 10, 64

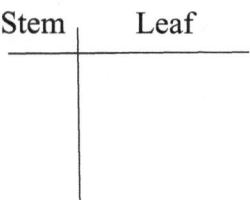

2) 43, 85, 52, 48, 45, 43, 51, 81, 59, 50, 85, 89

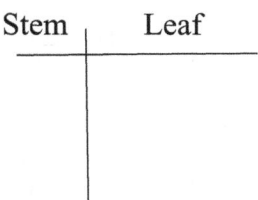

3) 112, 39, 46, 35, 80, 119, 42, 114, 37, 112, 47, 119

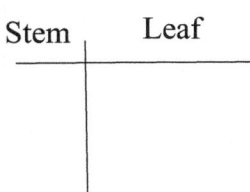

4) 90, 50, 131, 93, 112, 56, 139, 98, 115, 59, 98, 135, 111

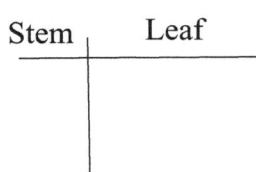

Scatter Plots

✎ Construct a scatter plot.

x	1	2	3	4	5	8
y	15	40	50	35	70	20

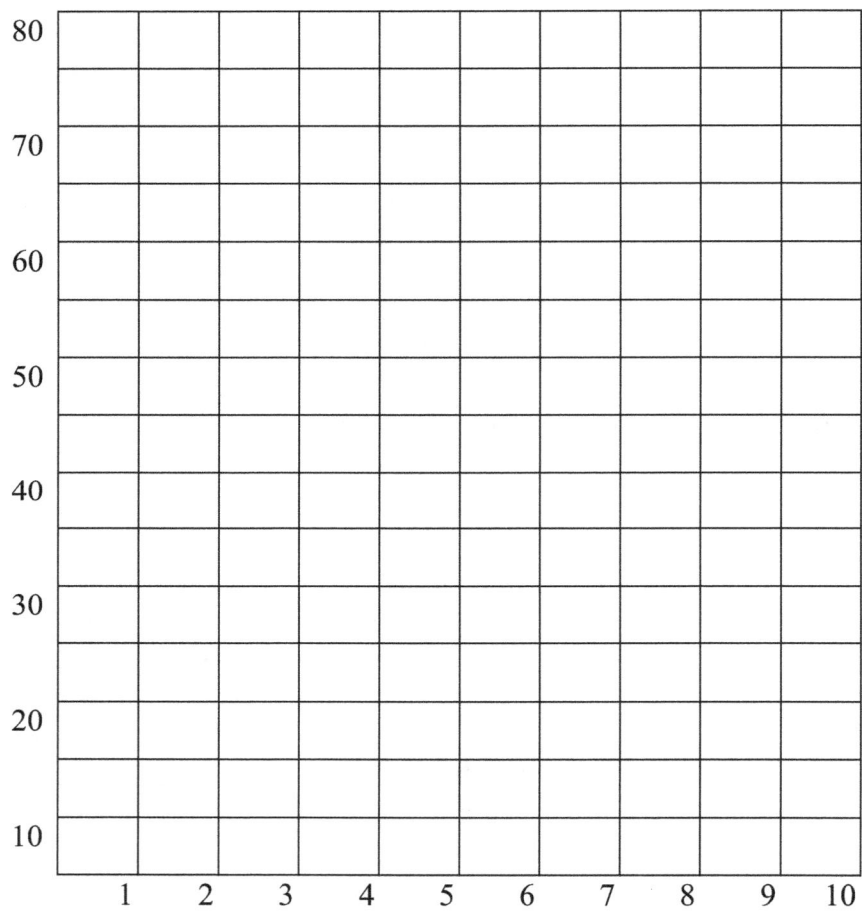

ISEE Lower-Level Subject Test Mathematics

Probability Problems

✏ Solve.

1) A number is chosen at random from 1 to 20. Find the probability of selecting a 10 or smaller.

2) A number is chosen at random from 1 to 25. Find the probability of selecting multiples of 5.

3) A number is chosen at random from 1 to 15. Find the probability of selecting multiples of 7.

4) A number is chosen at random from 1 to 20. Find the probability of selecting a multiple of 6.

5) A number is chosen at random from 1 to 10. Find the probability of selecting prime numbers.

6) A number is chosen at random from 1 to 20. Find the probability of not selecting factors of 15.

ISEE Lower-Level Subject Test Mathematics

Answers of Worksheets

Mean and Median

1) Mean: 13.4, Median: 13
2) Mean: 12.67, Median: 13
3) Mean: 18.6, Median: 14
4) Mean: 4.33, Median: 5
5) Mean: 12, Median: 11
6) Mean: 13.57, Median: 9
7) Mean: 20.125, Median: 18
8) Mean: 18.57, Median: 14
9) Mean: 35.43, Median: 32
10) Mean: 11.5, Median: 12
11) Mean: 34.67, Median: 38
12) Mean: 47, Median: 48
13) Mean: 48.33, Median: 42
14) Mean: 63.86, Median: 67
15) Mean: 66.33, Median: 69
16) Mean: 38.86, Median: 38
17) Mean: 67.4, Median: 67
18) Mean: 8.6, Median: 7

Mode and Range

1) Mode: 4, Range: 10
2) Mode: 8, Range: 39
3) Mode: 3, Range: 21
4) Mode: 4, Range: 35
5) Mode: 5, Range: 11
6) Mode: 7, Range: 17
7) Mode: 9, Range: 21
8) Mode: 11, Range: 41
9) Mode: 3, Range: 17
10) Mode: 14, Range: 29
11) Mode: 4, Range: 24
12) Mode: 7, Range: 29
13) Mode: 14, Range: 37
14) Mode: 14, Range: 22

Graph Points on a Coordinate Plane

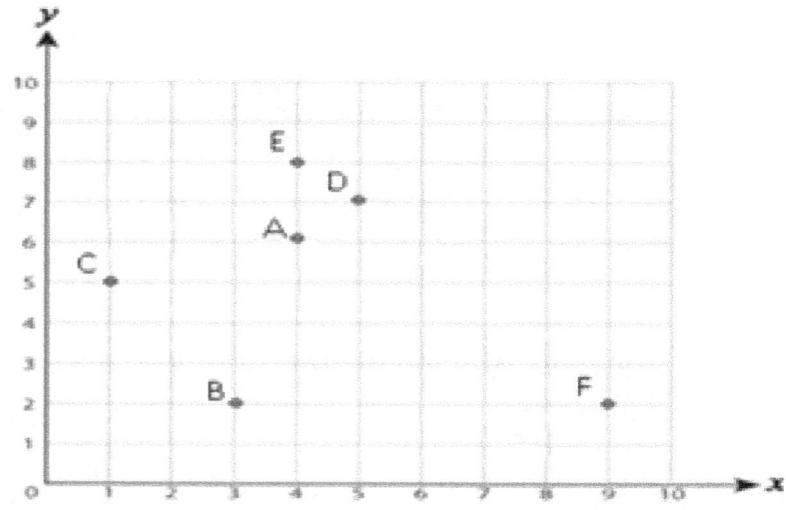

WWW.MathNotion.Com

ISEE Lower-Level Subject Test Mathematics

Bar Graph

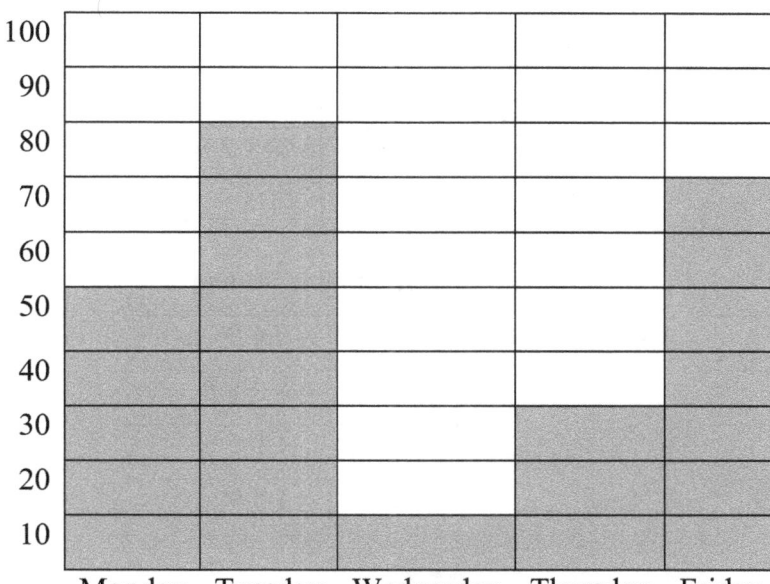

Tally and Pictographs

Dot plots

1) 22
2) 5
3) 11
4) 10 and 14
5) 2
6) 4
7) 2

Line Graphs

1) 6
2) Monday
3) Thursday
4) 30

ISEE Lower-Level Subject Test Mathematics

Stem–And–Leaf Plot

1)

Stem	leaf
1	0 4 9
4	2 2 2 7 9
6	4 5 9

2)

Stem	leaf
4	3 3 5 8
5	0 1 2 9
8	1 5 5 9

3)

Stem	leaf
3	5 7 9
4	2 6 7
8	0
11	2 2 4 9 9

4)

Stem	leaf
5	0 6 9
9	0 3 8 8
11	1 2 5
13	1 5 9

Scatter Plots

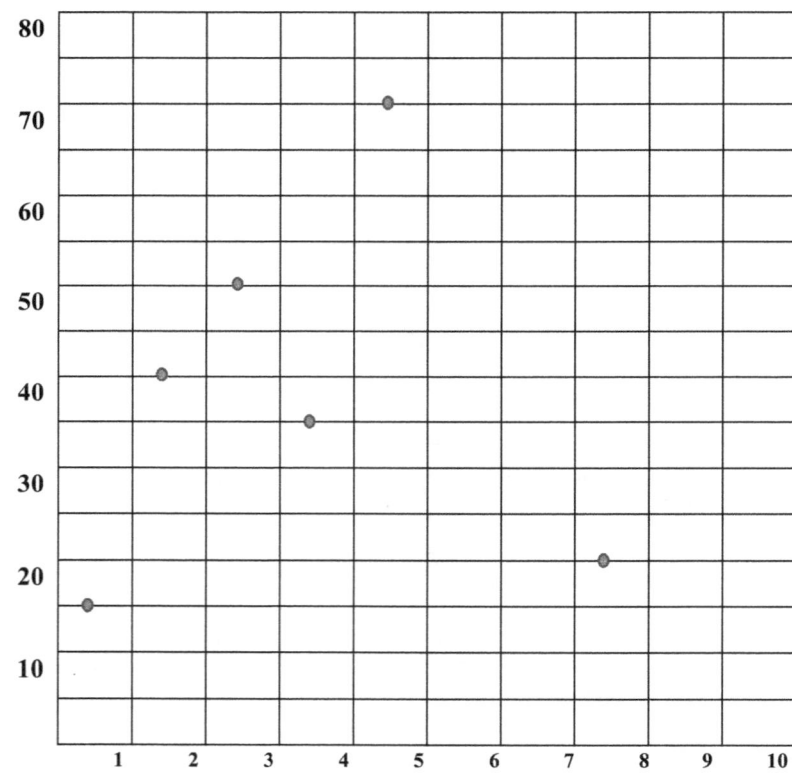

Probability Problems

1) $\frac{1}{2}$

2) $\frac{1}{5}$

3) $\frac{2}{15}$

4) $\frac{3}{20}$

5) $\frac{2}{5}$

6) $\frac{4}{5}$

ISEE Lower-Level Subject Test Mathematics

Chapter 13 : ISEE Lower-Level Practice Tests

The Independent School Entrance Exam (ISEE) is an admission test developed by the Educational Records Bureau for its member schools as part of their admission process.

ISEE Lower-Level tests use a multiple-choice format and contain two Mathematics sections:

Quantitative Reasoning:

There are 38 questions in the Quantitative Reasoning section and students have 35 minutes to answer the questions. This section contains word problems requiring either no calculation or simple calculation.

Mathematics Achievement:

There are 30 questions in the Mathematics Achievement section and students have 30 minutes to answer the questions. Mathematics Achievement measures students' knowledge of Mathematics requiring one or more steps in calculating the answer.

In this section, there are two complete ISEE Lower-Level Quantitative Reasoning and Mathematics Achievement Tests. Let your student take these tests to see what score they'll be able to receive on a real ISEE Lower-Level test.

ISEE Lower-Level Subject Test Mathematics

Time to Test

Time to refine your skill with a practice examination.

Take a practice ISEE Lower-Level Math Test to simulate the test day experience. After you've finished, score your test using the answer key.

Before You Start

- You'll need a pencil and scratch papers to take the test.
- For each question, there are four possible answers. Choose which one is best.
- It's okay to guess. You won't lose any points if you're wrong.
- Use the answer sheet provided to record your answers.
- After you've finished the test, review the answer key to see where you went wrong.
- **Calculators are NOT allowed for the ISEE Lower Level Test.**

Good Luck!

ISEE Lower-Level Subject Test Mathematics

ISEE Lower-Level Practice Test Answer Sheets

Remove (or photocopy) these answer sheets and use them to complete the practice tests.

ISEE Lower-Level Practice Test

Quantitative Reasoning

#		#	
1	Ⓐ Ⓑ Ⓒ Ⓓ	21	Ⓐ Ⓑ Ⓒ Ⓓ
2	Ⓐ Ⓑ Ⓒ Ⓓ	22	Ⓐ Ⓑ Ⓒ Ⓓ
3	Ⓐ Ⓑ Ⓒ Ⓓ	23	Ⓐ Ⓑ Ⓒ Ⓓ
4	Ⓐ Ⓑ Ⓒ Ⓓ	24	Ⓐ Ⓑ Ⓒ Ⓓ
5	Ⓐ Ⓑ Ⓒ Ⓓ	25	Ⓐ Ⓑ Ⓒ Ⓓ
6	Ⓐ Ⓑ Ⓒ Ⓓ	26	Ⓐ Ⓑ Ⓒ Ⓓ
7	Ⓐ Ⓑ Ⓒ Ⓓ	27	Ⓐ Ⓑ Ⓒ Ⓓ
8	Ⓐ Ⓑ Ⓒ Ⓓ	28	Ⓐ Ⓑ Ⓒ Ⓓ
9	Ⓐ Ⓑ Ⓒ Ⓓ	29	Ⓐ Ⓑ Ⓒ Ⓓ
10	Ⓐ Ⓑ Ⓒ Ⓓ	30	Ⓐ Ⓑ Ⓒ Ⓓ
11	Ⓐ Ⓑ Ⓒ Ⓓ	31	Ⓐ Ⓑ Ⓒ Ⓓ
12	Ⓐ Ⓑ Ⓒ Ⓓ	32	Ⓐ Ⓑ Ⓒ Ⓓ
13	Ⓐ Ⓑ Ⓒ Ⓓ	33	Ⓐ Ⓑ Ⓒ Ⓓ
14	Ⓐ Ⓑ Ⓒ Ⓓ	34	Ⓐ Ⓑ Ⓒ Ⓓ
15	Ⓐ Ⓑ Ⓒ Ⓓ	35	Ⓐ Ⓑ Ⓒ Ⓓ
16	Ⓐ Ⓑ Ⓒ Ⓓ	36	Ⓐ Ⓑ Ⓒ Ⓓ
17	Ⓐ Ⓑ Ⓒ Ⓓ	37	Ⓐ Ⓑ Ⓒ Ⓓ
18	Ⓐ Ⓑ Ⓒ Ⓓ	38	Ⓐ Ⓑ Ⓒ Ⓓ
19	Ⓐ Ⓑ Ⓒ Ⓓ	39	Ⓐ Ⓑ Ⓒ Ⓓ
20	Ⓐ Ⓑ Ⓒ Ⓓ	40	Ⓐ Ⓑ Ⓒ Ⓓ

Mathematics Achievement

#		#	
1	Ⓐ Ⓑ Ⓒ Ⓓ	21	Ⓐ Ⓑ Ⓒ Ⓓ
2	Ⓐ Ⓑ Ⓒ Ⓓ	22	Ⓐ Ⓑ Ⓒ Ⓓ
3	Ⓐ Ⓑ Ⓒ Ⓓ	23	Ⓐ Ⓑ Ⓒ Ⓓ
4	Ⓐ Ⓑ Ⓒ Ⓓ	24	Ⓐ Ⓑ Ⓒ Ⓓ
5	Ⓐ Ⓑ Ⓒ Ⓓ	25	Ⓐ Ⓑ Ⓒ Ⓓ
6	Ⓐ Ⓑ Ⓒ Ⓓ	26	Ⓐ Ⓑ Ⓒ Ⓓ
7	Ⓐ Ⓑ Ⓒ Ⓓ	27	Ⓐ Ⓑ Ⓒ Ⓓ
8	Ⓐ Ⓑ Ⓒ Ⓓ	28	Ⓐ Ⓑ Ⓒ Ⓓ
9	Ⓐ Ⓑ Ⓒ Ⓓ	29	Ⓐ Ⓑ Ⓒ Ⓓ
10	Ⓐ Ⓑ Ⓒ Ⓓ	30	Ⓐ Ⓑ Ⓒ Ⓓ
11	Ⓐ Ⓑ Ⓒ Ⓓ	31	Ⓐ Ⓑ Ⓒ Ⓓ
12	Ⓐ Ⓑ Ⓒ Ⓓ	32	Ⓐ Ⓑ Ⓒ Ⓓ
13	Ⓐ Ⓑ Ⓒ Ⓓ	33	Ⓐ Ⓑ Ⓒ Ⓓ
14	Ⓐ Ⓑ Ⓒ Ⓓ	34	Ⓐ Ⓑ Ⓒ Ⓓ
15	Ⓐ Ⓑ Ⓒ Ⓓ	35	Ⓐ Ⓑ Ⓒ Ⓓ
16	Ⓐ Ⓑ Ⓒ Ⓓ	36	Ⓐ Ⓑ Ⓒ Ⓓ
17	Ⓐ Ⓑ Ⓒ Ⓓ	37	Ⓐ Ⓑ Ⓒ Ⓓ
18	Ⓐ Ⓑ Ⓒ Ⓓ	38	Ⓐ Ⓑ Ⓒ Ⓓ
19	Ⓐ Ⓑ Ⓒ Ⓓ	39	Ⓐ Ⓑ Ⓒ Ⓓ
20	Ⓐ Ⓑ Ⓒ Ⓓ	40	Ⓐ Ⓑ Ⓒ Ⓓ

WWW.MathNotion.Com

ISEE Lower-Level Subject Test Mathematics

ISEE Lower-Level Practice Test 1

Mathematics

Quantitative Reasoning

- ❖ 38 Questions.
- ❖ Total time for this test: 35 Minutes.
- ❖ You may NOT use a calculator for this test.

Released *Month Year*

ISEE Lower-Level Subject Test Mathematics

1) If 5 added to a number, the sum is 16. If the same number added to 25, the answer is?

    A. 29

    B. 34

    C. 36

    D. 28

2) $\frac{6+4+5\times 3+5}{8+12} = ?$

    A. $\frac{5}{8}$

    B. $\frac{9}{8}$

    C. $\frac{3}{2}$

    D. $\frac{2}{3}$

3) $3 \times 6 \times 12 \times 7$ is equal to the product of 18 and

    A. 69

    B. 96

    C. 72

    D. 84

4) If 56 can be divided by both 7 and x without leaving a remainder, then 56 can also be divided by which of the following?

    A. $x + 9$

    B. $2x - 5$

    C. $x + 5$

    D. $x \times 2 - 2$

5) Use the equations below to answer the question:

$$x + 16 = 23$$

$$11 + y = 19$$

What is the value of $x + y$?

    A. 15

    B. 13

    C. 20

    D. 17

WWW.MathNotion.Com

ISEE Lower-Level Subject Test Mathematics

6) Which of the following expressions has the same value as $\frac{9}{4} \times \frac{8}{5}$?

   A. $\frac{4 \times 6}{5}$

   B. $\frac{7 \times 2}{10}$

   C. $\frac{9 \times 5}{4 \times 8}$

   D. $\frac{2 \times 9}{5}$

7) When 6 is added to five times number N, the result is 46. Then N is ….

   A. 7

   B. 8

   C. 30

   D. 40

8) At noon, the temperature was 31 degrees. By midnight, it had dropped another 37 degrees. What was the temperature at midnight?

   A. 31 degrees above zero

   B. 37 degrees below zero

   C. 6 degrees above zero

   D. 6 degrees below zero

9) If a triangle has a base of 5 cm and a height of 12 cm, what is the area of the triangle?

   A. 25 cm^2

   B. 30 cm^2

   C. 60 cm^2

   D. 15 cm^2

10) Which formula would you use to find the area of a square?

    A. $\frac{1}{2}$ Base × Height

    B. Length × Width × Height

    C. Length × Width

    D. Side × Side

WWW.MathNotion.Com

ISEE Lower-Level Subject Test Mathematics

11) What is the next number in this sequence?

    4, 7, 11, 16, 22, ...

    A. 29   C. 31

    B. 28   D. 33

12) What is the average of the following numbers?

    16, 20, 22, 30, 44, 54

    A. 29   C. 31

    B. 28.5   D. 30.5

13) If there are 8 red balls and 14 blue balls in a basket, what is the probability that John will pick out a red ball from the basket?

    A. $\frac{7}{11}$ C. $\frac{8}{14}$

    B. $\frac{4}{11}$ D. $\frac{5}{14}$

14) How many lines of symmetry does an equilateral rectangle have?

    A. 6   C. 2

    B. 1   D. 4

15) What is %25 of 620?

    A. 95   C. 155

    B. 75   D. 105

ISEE Lower-Level Subject Test Mathematics

16) Which of the following statement is False?

   A. $63 \times \frac{1}{7} = 9$

   B. $(7 + 4) \times 3 = 33$

   C. $8 \div (4 - 2) = 0$

   D. $6 \times (8 - 6) = 12$

17) If all the sides in the following figure are of equal length and length of one side is 4, what is the perimeter of the figure?

   A. 16

   B. 36

   C. 24

   D. 44

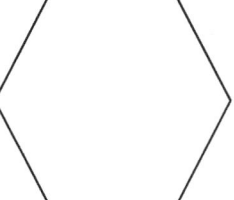

18) $\frac{9}{10} - \frac{7}{10} = ?$

   A. 0.8

   B. 0.02

   C. 0.20

   D. 0.6

19) If $N = 9$ and $\frac{18}{N} + 10 = \square$, then $\square = \ldots$

   A. 8

   B. 19

   C. 10

   D. 12

20) Three people can paint 3 houses in 6 days. How many people are needed to paint 4 houses in 2 days?

   A. 6

   B. 18

   C. 12

   D. 24

ISEE Lower-Level Subject Test Mathematics

21) Which of the following is greater than $\frac{17}{25}$?

A. $\frac{10}{16}$

B. $\frac{7}{8}$

C. $\frac{30}{50}$

D. 0.15

The result of a research shows the number of men and women in four cities of a country.

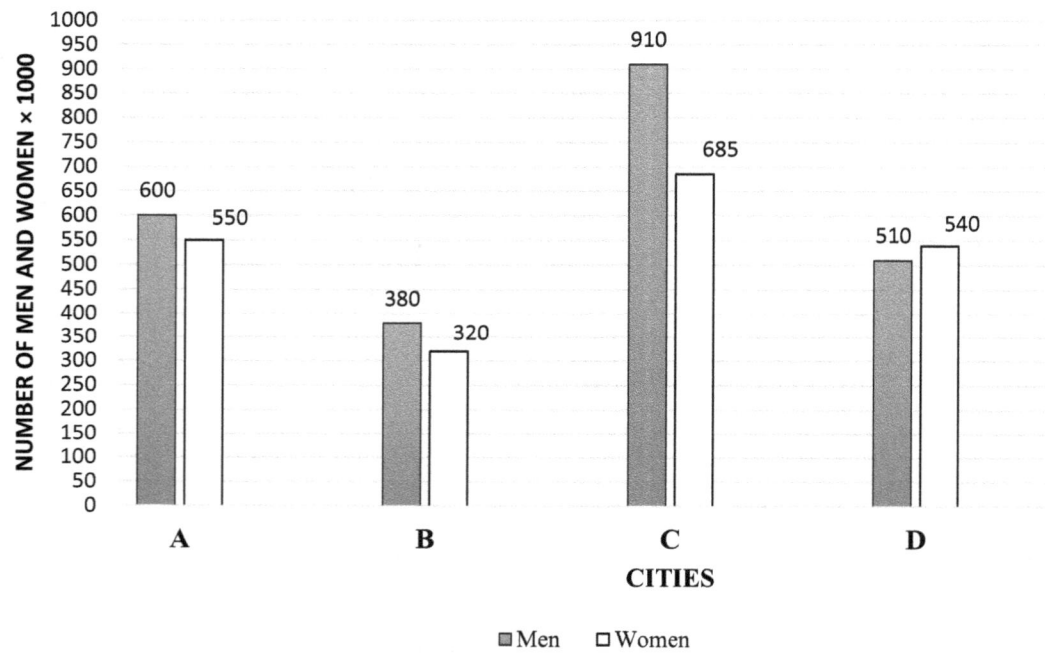

22) What is the difference of the population of men in the biggest city and in the smallest city?

A. 480

B. 610

C. 530

D. 270

ISEE Lower-Level Subject Test Mathematics

23) 8.9 − 5.38 is closest to which of the following.

   A. 5.2

   B. 3.5

   C. 5.5

   D. 3.6

24) Numbers x and y are shown below. How many times larger is the value of digit 9 in the number x, than the value of digit 9 in the number y?

$$x = 1{,}389 \quad y = 759$$

   A. 1

   B. 10

   C. 100

   D. 1,000

25) What is 3,685.77947 rounded to the nearest tenth?

   A. 3,685.78

   B. 3,685.80

   C. 3,686

   D. 3,685.779

26) $7a + 17 = 59, a = ?$

   A. 6

   B. 7

   C. 11

   D. 9

27) Two angles of a triangle measure 43 and 92. What is the measure of third angle?

   D. 65

   B. 135

   C. 55

   D. 45

ISEE Lower-Level Subject Test Mathematics

28) A woman weighs 147 pounds. She gains 52 pounds one month and 14 pounds the next month. What is her new weight?

   A. 223 Pounds

   B. 203 Pounds

   C. 213 Pounds

   D. 198 Pounds

29) If $\frac{1}{9}$ of a number is greater than 6, the number must be ……

   A. Less than 9

   B. Less than 36

   C. Less than 54

   D. Greater than 54

30) If $3 \times (M + N) = 27$ and M and N are whole numbers, then N could Not be?

   A. 0

   B. 7

   C. 4

   D. 12

31) At a Zoo, the ratio of lions to tigers is 28 to 12. Which of the following could NOT be the total number of lions and tigers in the zoo?

   A. 20

   B. 90

   C. 95

   D. 40

32) In the multiplication bellow, A represents which digit?

$$14 \times 3A2 = 5,348$$

   A. 9

   B. 4

   C. 8

   D. 6

33) If N is an even number, which of the following is always an odd number?

   A. $5N$

   B. $N + 10$

   C. $\frac{N}{2}$

   D. $N + 3$

ISEE Lower-Level Subject Test Mathematics

34) In a basket, there are equal numbers of red, black, yellow, and purple cards.

   Which of the following could be the number of cards in the basket?

   A. 27

   B. 181

   C. 58

   D. 92

35) Jim types 75 words per minute. How many words does he type in 20 seconds?

   A. 50

   B. 20

   C. 35

   D. 25

36) Which of the following is NOT equal to $\frac{5}{9}$?

   A. $\frac{25}{45}$

   B. $\frac{45}{81}$

   C. $\frac{30}{54}$

   D. $\frac{15}{36}$

37) Which of the following is closest to 7.01?

   A. 8

   B. 7.1

   C. 7

   D. 7.11

38) What is the median of these numbers? 5, 11, 18, 15, 20, 25, 3

   A. 20

   B. 22.5

   C. 15

   D. 18

ISEE Lower-Level Subject Test Mathematics

ISEE Lower-Level Subject Test Mathematics

ISEE Lower-Level Practice Test 1

Mathematics

Mathematics Achievement

- ❖ 30 Questions.
- ❖ Total time for this test: 30 Minutes.
- ❖ You may NOT use a calculator for this test.

Released *Month Year*

ISEE Lower-Level Subject Test Mathematics

1) What is 6,682.3749 rounded to the nearest tenth?

   A. 6,682.40

   B. 6,682.395

   C. 6,683

   D. 6,682.3800

2) Which of the following fractions is the largest?

   A. $\frac{2}{7}$

   B. $\frac{1}{5}$

   C. $\frac{8}{10}$

   D. $\frac{6}{10}$

3) A bag contains 16 balls: two green, two black, five blue, a brown, three red and three white. If 15 balls are removed from the bag at random, what is the probability that a brown ball has been removed?

   A. $\frac{1}{15}$

   B. $\frac{5}{8}$

   C. $\frac{1}{16}$

   D. $\frac{15}{16}$

4) From last year, the price of gasoline has increased from $1.35 per gallon to $1.89 per gallon. The new price is what percent of the original price?

   A. 124%

   B. 140%

   C. 135%

   D. 150%

5) Emma purchased a computer for $196. The computer is regularly priced at $280. What was the percent discount Emma received on the computer?

   A. 35%

   B. 65%

   C. 70%

   D. 30%

ISEE Lower-Level Subject Test Mathematics

6) In the given diagram, the height is 15 cm. what is the area of the triangle?

   A. 90 cm²

   B. 120 cm²

   C. 180 cm²

   D. 360 cm²

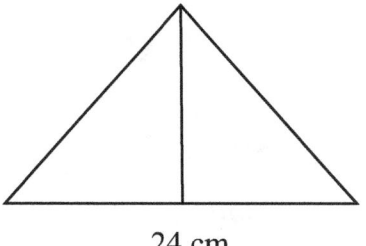

24 cm

7) Two angles of a triangle measure 46 and 87. What is the measure of the third angle?

   A. 47° C. 113°

   B. 45° D. 133°

8) $\frac{1}{2} + \frac{2}{9} =$

   A. $\frac{11}{18}$ C. $\frac{3}{11}$

   B. $\frac{3}{10}$ D. $\frac{13}{18}$

9) What is the least common multiple (LCM) of 8 and 17?

   A. Have no common multiples. C. 17

   B. 156 D. 136

10) While at work, Emma checks her email once every 150 minutes. In 5-hour, how many times does she check her email?

    A. 4 Times C. 10 Times

    B. 5 Times D. 2 Times

WWW.MathNotion.Com

ISEE Lower-Level Subject Test Mathematics

Use the following table to answer question below.

DANIEL'S BIRD-WATCHING PROJECT

DAY	NUMBER OF RAPTORS SEEN
Monday	?
Tuesday	17
Wednesday	20
Thursday	15
Friday	8
MEAN	15

11) This table shows the data Daniel collects while watching birds for one week. How many raptors did Daniel see on Monday?

A. 15
B. 14
C. 18
D. 13

12) Which of the following is NOT a factor of 32?

A. 8
B. 2
C. 64
D. 16

13) If a rectangular swimming pool has a perimeter of 52 feet and is 9 feet wide, what is its area?

A. 153 ft
B. 183 ft
C. 135 ft
D. 17 ft

ISEE Lower-Level Subject Test Mathematics

14) Which of the following is an obtuse angle?

   A. 135°

   B. 65°

   C. 89°

   D. 190°

15) In the following figure, the shaded squares are what fractional part of the whole set of squares?

   A. $\frac{1}{6}$

   B. $\frac{3}{8}$

   C. $\frac{3}{10}$

   D. $\frac{1}{3}$

16) If a box contains red and blue balls in ratio of 2:3 red to blue, how many red balls are there if 72 blue balls are in the box?

   A. 52

   B. 48

   C. 67

   D. 44

17) A shirt costing $700 is discounted 22%. After a month, the shirt is discounted another 13%. Which of the following expressions can be used to find the selling price of the shirt?

   A. (700)(0.87)

   B. (700) − 700(0.78)

   C. (700)(0.87) − (700)(0.78)

   D. (700)(0.78)(0.87)

ISEE Lower-Level Subject Test Mathematics

18) Emma draws a shape on her paper. The shape has four sides. It has only one pair of parallel sides. What shape does Emma draw?

   A. Parallelogram   C. Square

   B. Rectangle   D. Trapezoid

19) If $A = 80$, then which of the following equations are correct?

   A. $2A + 40 = 120 + A$ C. $40 \times A = 120 \div A$

   B. $A \div 40 = 120 - 2A$ D. $A - 40 = 120 + A$

20) Joe makes $4.80 per hour at his work. If he works 10 hours, how much money will he earn?

   A. $96   C. $48

   B. $84.80   D. $48.80

21) In a classroom of 60 students, 18 are male. About what percentage of the class is female?

   A. 70%   C. 78%

   B. 30%   D. 35%

22) Nancy ordered 12 pizzas. Each pizza has 8 slices. How many slices of pizza did Nancy ordered?

   A. 80   C. 69

   B. 96   D. 120

ISEE Lower-Level Subject Test Mathematics

23) Mike is 21 miles ahead of Julia running at 5.5 miles per hour and Julia is running at the speed of 9 miles per hour. How long does it take Julia to catch Mike?

   A. 3 hours

   B. 3.5 hours

   C. 4.5 hours

   D. 4 hours

24) Convert 0.079 to a percent.

   A. 0.079%

   B. 0.79%

   C. 7.90%

   D. 79%

25) Julie gives 6 pieces of candy to each of her friends. If Julie gives all her candy away, which amount of candy could have been the amount she distributed?

   A. 74

   B. 336

   C. 188

   D. 227

26) A taxi driver earns $7 per 1-hour work. If he works 10 hours a day and in 1 hour, he uses 3-liters petrol with price $1.4 for 1-liter. How much money does he earn in one day?

   A. $28

   B. $15

   C. $55

   D. $30

27) The number 0.003 can also represent by which of the following?

   A. $\frac{3}{100}$

   B. $\frac{3}{1,000}$

   C. $\frac{3}{10,000}$

   D. $\frac{3}{10}$

ISEE Lower-Level Subject Test Mathematics

28) 23 hr. 10 min.
    −15 hr. 38 min.
    ─────────────

A. 8 hr. 22 min.

B. 8 hr. 32 min.

C. 7 hr. 32 min.

D. 5 hr. 22 min.

29) 160 students took an exam and 48 of them failed. What percent of the students passed the exam?

A. 15 %

B. 38 %

C. 30 %

D. 70 %

30) The width of a box is one third of its length. The height of the box is one eighth of its width. If the length of the box is 24 cm, what is the volume of the box?

A. 82 cm^3

B. 99 cm^3

C. 182 cm^3

D. 192 cm^3

STOP

IF YOU FINISH BEFORE TIME IS CALLED, YOU MAY CHECK YOUR WORK ON THIS SECTION ONLY. DO NOT TURN TO ANY OTHER SECTION IN THE TEST.

ISEE Lower-Level Practice Test 2

Mathematics

Quantitative Reasoning

- ❖ 38 Questions.
- ❖ Total time for this test: 35 Minutes.
- ❖ You may NOT use a calculator for this test.

Released *Month Year*

ISEE Lower-Level Subject Test Mathematics

1) Find the missing number in the sequence: 10, 12, 15, …., 24.

   A. 18

   B. 19

   C. 21

   D. 22

2) The length of a rectangle is 7 times of its width. If the length is 28, what is the perimeter of the rectangle?

   A. 55

   B. 38

   C. 64

   D. 72

3) Mary has y dollars. John has $17 more than Mary. If John gives Mary $30, then in terms of y, how much does John have now?

   A. $y + 13$

   B. $y + 23$

   C. $y - 13$

   D. $y - 23$

4) Dividing 193 by 6 leaves a remainder of

   A. 5

   B. 4

   C. 2

   D. 1

5) If $9,000 + A - 4,700 = 8,200$, then $A =$

   A. 900

   B. 3,900

   C. 3,500

   D. 2,400

6) When 52 is divided by 3, the remainder is the same as when 25 is divided by?

   A. 9

   B. 7

   C. 9

   D. 6

WWW.MathNotion.Com

ISEE Lower-Level Subject Test Mathematics

7) For what price is 54 percent off the same as $270 off?

   A. $500

   B. $400

   C. $600

   D. $700

8) Which of the following fractions is less than $\frac{9}{5}$?

   A. $\frac{169}{100}$

   B. $\frac{15}{4}$

   C. $\frac{31.2}{10}$

   D. $\frac{26}{10}$

9) Use the equation below to answer the question.

$$x + 7 = 12$$

$$7y = 63$$

   What is the value of $y + x$?

   A. 4

   B. 6

   C. 16

   D. 14

10) If $638 - x + 250 = 518$, then $x = ?$

   A. 350

   B. 460

   C. 370

   D. 420

11) In the following right triangle, what is the value of x?

   A. 60

   B. 50

   C. 45

   D. 37

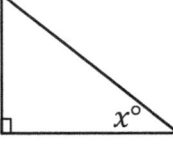

ISEE Lower-Level Subject Test Mathematics

12) Of the following, 50 percent of $63.78 is closest to?

   A. $14.00

   B. $28.00

   C. $24.00

   D. $32.00

13) Solve: 14.58 − 9.6 = ...?

   A. 5.68

   B. 4.88

   C. 4.98

   D. 5.12

14) $\frac{17}{9} - \frac{8}{9} = ?$

   A. 1

   B. 1.2

   C. 9.1

   D. 1.9

15) When 5 is added to seven times a number N, the result is 52. Which of the following equations represents this statement?

   A. $7 + 5N = 52$

   B. $52N + 5 = 7$

   C. $7N + 5 = 52$

   D. $7N + 52 = 5$

16) If $82 = 5 \times Z + 17$, then $Z = $ ....

   A. 9

   B. 13

   C. 19

   D. 15

17) John has 3,700 cards and Max has 859 cards. How many more cards does John have than Max?

   A. 2,841

   B. 2,541

   C. 2,632

   D. 2,532

ISEE Lower-Level Subject Test Mathematics

18) What is 4 percent of 650?

   A. 35

   B. 26

   C. 42

   D. 28

19) Which of the following statements is False?

   A. $(7 \times 2 + 6) \times 7 = 140$

   B. $(2 \times 8 + 9) \div 5 = 5$

   C. $12 + (5 \times 6) = 42$

   D. $75 \div (15 + 10) = 13$

20) The distance between cities A and B is approximately 4,200 miles. If Nicole drives an average of 79 miles per hour, how many hours will it take her to drive from city A to city B?

   A. approximately 47 hours

   B. approximately 49 hours

   C. approximately 54 hours

   D. approximately 53 hours

21) $\frac{56}{70}$ is equal to:

   A. 8.5

   B. 0.85

   C. 0.8

   D. 0.08

22) Which of the following is NOT a prime factor of 84?

   A. 3

   B. 7

   C. 2

   D. 4

23) A writer finishes 240 pages of his manuscript in 60 hours. How many pages is his average?

   A. 18

   B. 4

   C. 8

   D. 24

ISEE Lower-Level Subject Test Mathematics

24) During the six-month period shown, what is the median number of shoes per month?

A. 40

B. 45

C. 50

D. 45.5

25) A trash container, when empty, weighs 47 pounds. If this container is filled with a load of trash that weighs 390 pounds, what is the total weight of the container and its contents?

A. 344 pounds

B. 437 pounds

C. 372 pounds

D. 472 pounds

26) $11a + 30 = 151, a = ?$

A. 11

B. 22

C. 16

D. 12

ISEE Lower-Level Subject Test Mathematics

27) What is the place value of 7 in 4.5871?

   A. Hundredths   C. Hundred thousandths

   B. Thousandths   D. Ten thousandths

28) Which of these numbers is equal to $\frac{95}{100,000}$?

   A. 0.95000   C. 0.00095

   B. 0.09500   D. 0.00950

29) During a 24-hour day, Moe works $\frac{1}{8}$ of the time. How many hours does Moe work in that day?

   A. 8   C. 3

   B. 6   D. 4

30) In a basket, the ratio of red marbles to blue marbles is 3 to 2. Which of the following could NOT be the total number of red and blue marbles in the basket?

   A. 15   C. 48

   B. 50   D. 30

31) A square has an area of $16 cm^2$. What is its perimeter?

   A. 32 cm   C. 8 cm

   B. 16 cm   D. 54 cm

32) If $300 + \square - 220 = 1,750$, then $\square = ?$

   A. 970   C. 1,667

   B. 1,260   D. 1,670

ISEE Lower-Level Subject Test Mathematics

33) There are 85 students in a class. If the ratio of the number of girls to the total number of students in the class is $\frac{1}{5}$, which are the following is the number of boys in that class?

    A. 17   C. 68

    B. 56   D. 45

34) If $N \times (7 - 4) = 48$ then $N =$?

    A. 16   C. 8

    B. 12   D. 24

35) If $x \blacksquare y = 2x + 5y - 12$, what is the value of $2 \blacksquare 6$?

    A. 24   C. 32

    B. 22   D. 28

36) Of the following, which number if the greatest?

    A. 0.0872   C. 0.7872

    B. 0.7845   D. 0.7827

37) $\frac{7}{8} - \frac{3}{4} = ?$

    A. 0.125   C. 0.4

    B. 0.25   D. 0.44

38) Which of the following is the closest to 4.03?

    A. 4.1   C. 4.3

    B. 4   D. 4.1

ISEE Lower-Level Practice Test 2

Mathematics

Mathematics Achievement

- 30 Questions.
- Total time for this test: 30 Minutes.
- You may NOT use a calculator for this test.

Released *Month Year*

ISEE Lower-Level Subject Test Mathematics

1) What's the greatest common factor of the 16 and 38?

   A. 19

   B. 16

   C. 2

   D. 8

2) Which is sixty-nine thousand, two hundred fourteen?

   A. 69,214

   B. 690,214

   C. 609,214

   D. 692,014

3) What is the name of a rectangle with sides of equal length?

   A. Hexagon

   B. Octagon

   C. Pentagon

   D. Square

4) With what number must 3.564852 be multiplied in order to obtain the number 35,648.52?

   A. 1000,000

   B. 1,000

   C. 10,000

   D. 100,000

5) A right triangle has two short sides with lengths 12 and 9. What is the perimeter of the triangle?

   A. 34

   B. 36

   C. 54

   D. 72

6) Which expression has a value of − 13?

   A. 10− (−2) + (−25)

   B. −8 + (−5) × (−2)

   C. (−4) × (−6) + (−2) × (−7)

   D. (−5) × (−9) + 20

ISEE Lower-Level Subject Test Mathematics

7) Lily and Ella are in a pancake–eating contest. Lily can eat nine pancakes per minute, while Ella can eat $4\frac{1}{2}$ pancakes per minute. How many total pancakes can they eat in 4 minutes?

   A. 37.5 Pancakes

   B. 56 Pancakes

   C. 54 Pancakes

   D. 49 Pancakes

8) $0.72 + 2.2 + 3.68 = ?$

   A. 4.4

   B. 6.88

   C. 6.6

   D. 5.8

9) What is the perimeter of a rectangle that has a length of 7 inches and a width of 5 inches?

   A. 12

   B. 24

   C. 35

   D. 28

10) How many $\frac{1}{10}$ cup servings are in a package of cheese that contains $4\frac{2}{5}$ cups altogether?

    A. 36

    B. 46

    C. 34

    D. 44

11) Which expression is equal to $\frac{7}{15}$?

    A. $7 - 15$

    B. $7 \div 15$

    C. 7×15

    D. $\frac{15}{7}$

ISEE Lower-Level Subject Test Mathematics

12) If the following clock shows a time in the morning, what time was it 6 hours and 30 minutes ago?

A. 11:15 AM

B. 11:45 AM

C. 10:45 PM

D. 10:15 PM

13) Which of the following is not a multiple of 6?

A. 116

B. 246

C. 96

D. 156

14) The area of a rectangle is 84 square meters. The width is 7 meters. What is the length of the rectangle?

A. 9

B. 16

C. 12

D. 14

15) The temperature on Sunday at 12:00 PM was 79°F. Low temperature on the same day was 47°F cooler. Which temperature is closest to the low temperature on that day?

A. 36°F

B. 32°F

C. 22°F

D. 63°F

16) $(8^2 - 5^2) \div (3^3 \div 3^2) =$ ___

A. $\frac{8}{3}$

B. 13

C. 12

D. $\frac{5}{21}$

WWW.MathNotion.Com

ISEE Lower-Level Subject Test Mathematics

Use the table below to answer the question.

City Populations	
City	Population
Denton	22,856
Bomberg	26,753
Windham	26,821
Sanhill	19,640

17) Which list of city populations is in order from least to greatest?

A. 22,856; 26,753; 26,821; 19,640

B. 26,821; 26,753; 22,856; 19,640

C. 19,640; 22,856; 26,753; 26,821

D. 19,640; 26,821; 26,753; 22,856

18) Ella buys six items costing $6.18, $8.13, $4.20, $6.15, $9.95, and $11.85. What is the estimated total cost of Ella's items?

A. between $32 and $38

B. between $20 and $28

C. between $44 and $48

D. between $28 and $32

19) How long is the line segment shown on the number line below?

A. 5

B. 10

C. 15

D. 6

ISEE Lower-Level Subject Test Mathematics

20) Which fraction has the least value?

   A. $\frac{5}{6}$

   B. $\frac{3}{8}$

   C. $\frac{3}{4}$

   D. $\frac{7}{24}$

21) What fraction of each shape is shaded?

   a) 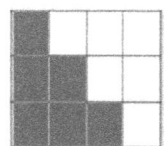   b)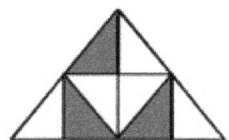

   A. a. $\frac{3}{10}$; b. $\frac{1}{8}$

   B. a. $\frac{1}{6}$; b. $\frac{5}{8}$

   C. a. $\frac{1}{2}$; b. $\frac{3}{8}$

   D. a. $\frac{1}{6}$; b. $\frac{3}{8}$

22) Which statement about the number 894,913.75 is true?

   A. The digit 7 has a value of (7 × 10)

   B. The digit 5 has a value of (5 × 10)

   C. The digit 9 has a value of (9 × 100)

   D. The digit 1 has a value of (1 × 100)

23) Elise described a number using these clues:

   Three – digit even numbers that have a 6 in the hundreds place and a 9 in the tens place.

   Which number could fit Elise's description?

   A. 963

   B. 694

   C. 966

   D. 693

ISEE Lower-Level Subject Test Mathematics

24) Jason's favorite sports team has won 0.55 of its games this season. How can Jason express this decimal as a fraction?

A. $\frac{45}{55}$

B. $\frac{100}{55}$

C. $\frac{11}{20}$

D. $\frac{15}{25}$

25) The following graph shows the mark of six students in mathematics. What is the mean (average) of the marks?

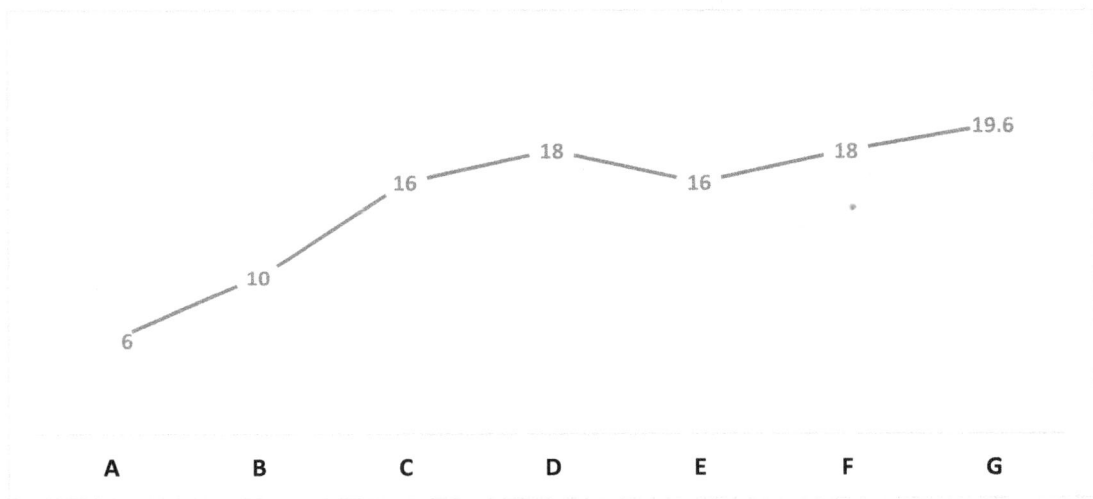

A. 13.9

B. 14.8

C. 14.6

D. 16

26) In the deck of cards, there are 5 spades, 6 hearts, 11 clubs, and 10 diamonds. What is the probability that William will pick out a spade?

A. $\frac{5}{32}$

B. $\frac{27}{32}$

C. $\frac{4}{5}$

D. $\frac{5}{27}$

ISEE Lower-Level Subject Test Mathematics

27) There are 66 students from Riddle Elementary school at the library on Monday. The other 17 students in the school are practicing in the classroom. Which number sentence shows the total number of students in Riddle Elementary school?

A. $66 + 17$

B. $66 - 17$

C. 66×17

D. $66 \div 17$

28) There are 365 days in a year, and 24 hours in a day. How many hours are in three years?

A. 8,820

B. 12,420

C. 16,880

D. 26,280

29) $\frac{19}{31}$ is equal to:

A. 0.61

B. 0.88

C. 0.125

D. 610

30) Which correctly shows the way to find $30 \times 18 = ?$

A. $10 \times 10 + 3 \times 8$

B. $30 \times 10 + 30 \times 8$

C. $30 \times 10 + 8$

D. $3 \times 18 + 10 \times 18$

STOP

IF YOU FINISH BEFORE TIME IS CALLED, YOU MAY CHECK YOUR WORK ON THIS SECTION ONLY. DO NOT TURN TO ANY OTHER SECTION IN THE TEST.

ISEE Lower-Level Subject Test Mathematics

Chapter 14 : Answers and Explanations
ISEE Lower-Level Practice Tests
Answer Key

❋ Now, it's time to review your results to see where you went wrong and what areas you need to improve!

ISEE Lower-Level Practice Test 1 - Mathematics

Quantitative Reasoning						Mathematics Achievement					
1	C	16	C	31	C	1	A	16	B		
2	C	17	C	32	C	2	C	17	D		
3	D	18	C	33	D	3	D	18	D		
4	D	19	D	34	D	4	B	19	A		
5	A	20	C	35	D	5	D	20	C		
6	D	21	B	36	D	6	C	21	A		
7	B	22	C	37	C	7	A	22	B		
8	D	23	B	38	C	8	D	23	D		
9	B	24	A			9	D	24	C		
10	D	25	B			10	D	25	B		
11	A	26	A			11	A	26	A		
12	C	27	D			12	C	27	B		
13	B	28	C			13	A	28	C		
14	C	29	D			14	A	29	D		
15	C	30	D			15	D	30	D		

WWW.MathNotion.Com

ISEE Lower-Level Subject Test Mathematics

ISEE Lower-Level Practice Test 2 - Mathematics

Quantitative Reasoning

1	B	16	B	31	B
2	C	17	A	32	D
3	C	18	B	33	C
4	D	19	D	34	A
5	C	20	D	35	B
6	D	21	C	36	C
7	A	22	D	37	A
8	A	23	B	38	B
9	D	24	B		
10	C	25	B		
11	C	26	A		
12	D	27	B		
13	C	28	C		
14	A	29	C		
15	C	30	C		

Mathematics Achievement

1	C	16	B
2	A	17	C
3	D	18	C
4	C	19	B
5	B	20	D
6	A	21	C
7	C	22	C
8	C	23	B
9	B	24	C
10	D	25	B
11	B	26	A
12	C	27	A
13	A	28	D
14	C	29	A
15	B	30	B

ISEE Lower-Level Subject Test Mathematics

Practice Tests 1:
Quantitative Reasoning

1) Answer: C.

Let x be the number. Then:

$5 + x = 16 \to x = 11 \to 11 + 25 = 36$

2) Answer: C.

$\frac{6+4+5\times 3+5}{8+12} = \frac{30}{20} = \frac{3}{2}$

3) Answer: D.

$3 \times 6 \times 12 \times 7$ is equal to the product of 18 and 84.

$(3 \times 6) \times (12 \times 7) = 18 \times 84$

4) Answer: D.

$56 = x \times 7 \to x = 56 \div 7 = 8$

x equals to 8. Let's review the options provided:

A) $x + 9 \to 8 + 9 = 17$ 56 is not divisible by 17.

B) $2x - 5 \to 2 \times 8 - 5 = 11$ 56 is not divisible by 11.

C) $x + 5 \to 8 + 5 = 13$ 56 is not divisible by 13.

D) $x \times 2 - 2 \to 8 \times 2 - 2 = 14$ 56 is divisible by 14.

The answer is D.

5) Answer: A.

$x + 16 = 23$; Subtract 16 from both sides $\to x + 16 - 16 = 23 - 16 \to x = 7$

$11 + y = 19$; Subtract 11 from both sides $\to 11 + y - 11 = 19 - 11 \to y = 8$

$x + y = 7 + 8 = 15$

6) Answer: D.

$\frac{9}{4} \times \frac{8}{5} = \frac{72}{20} = \frac{18}{5}$

Choice D is equal to $\frac{18}{5}$. $\frac{2 \times 9}{5} = \frac{18}{5}$

7) Answer: B.

$6 + 5N = 46 \to 5N = 46 - 6 \to 5N = 40 \to N = 8$

ISEE Lower-Level Subject Test Mathematics

8) Answer: D.

$31 - 37 = -6$. The temperature at midnight was 6 degrees below zero.

9) Answer: B.

Area of a triangle $= \frac{1}{2} \times (base) \times (height) = \frac{1}{2} \times 5 \times 12 = 30$

10) Answer: D.

$area\ of\ a\ square = side \times side$

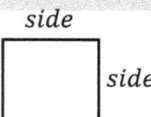

11) Answer: A.

First, find the pattern,

$4 + 3 = 7 \rightarrow 7 + 4 = 11 \rightarrow 11 + 5 = 16 \rightarrow 16 + 6 = 22$

The difference of two consecutive numbers increases by 1. The difference of 16 and 22 is 6. So, the next number should be 29. $22 + 7 = 29$

12) Answer: C.

average $= \frac{sum\ of\ all\ numbers}{number\ of\ numbers} = \frac{16+20+22+30+44+54}{6} = 31$

13) Answer: B.

There are 8 red ball and 22 are total number of balls.

Therefore, probability that John will pick out a red ball from the basket is 8 out of 22 or $\frac{8}{8+14} = \frac{8}{22} = \frac{4}{11}$.

14) Answer: C.

An equilateral rectangle has 2 lines of symmetry.

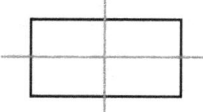

15) Answer: C.

25 percent of $620 = 25\%$ of $620 = \frac{25}{100} \times 620 = 155$

16) Answer: C.

Let's review the options provided:

A. $63 \times \frac{1}{7} = \frac{63}{7} = 9$ This is true!

B. $(7 + 4) \times 3 = 33$ This is true!

ISEE Lower-Level Subject Test Mathematics

C. $8 \div (4 - 2) = 0 \rightarrow 8 \div 2 = 4$ This is NOT true!

D. $6 \times (8 - 6) = 12 \rightarrow 6 \times 2 = 12$ This is true!

17) Answer: C.

The shape has 6 equal sides. And its side is 4. Then, the perimeter of the shape is:

$4 \times 6 = 24$

18) Answer: C.

$\frac{9}{10} - \frac{7}{10} = \frac{2}{10} = 0.2 = 0.20$

19) Answer: D.

$N = 9$ then: $\frac{18}{9} + 10 = 2 + 10 = 12$

20) Answer: C.

Three people can paint 3 houses in 6 days. $3 \times 6 = 18$ man days to paint 3 houses, 6-man days per house ($18 \div 3 = 6$).

$6 \times 4 = 24$ man-days for 6 houses; $\frac{24}{2} = 12$ painters.

21) Answer: B.

$\frac{17}{25} = 0.68$, the only choice that is greater than 0.68 is $\frac{7}{8}$.

$\frac{7}{8} = 0.875$, $0.875 > 0.68$

22) Answer: C.

The biggest city is city C, and the smallest city is city B.

Number of men in city C is 910 and number of men in city B is 380.

Then: $910 - 380 = 530$

23) Answer: B.

$8.9 - 5.38 = 3.52$, which is closest to 3.5.

24) Answer: A.

The value of digit 9 in both numbers x and y are in the ones place. Therefore, they have the same value.

25) Answer: B.

3,685.77957 rounded to the nearest tenth is 3,685.8.

ISEE Lower-Level Subject Test Mathematics

26) Answer: A.

$7a + 17 = 59 \to 7a = 59 - 17$

$7a = 42 \to a = 6$

27) Answer: D.

All angles in a triangle sum up to 180 degrees.

Two angles of a triangle measure 43 and 92.

$43 + 92 = 135$. Then, the third angle is: $180 - 135 = 45$

28) Answer: C.

$147 + 52 + 14 = 213$ Pounds

29) Answer: D.

If $\frac{1}{9}$ of a number is greater than 6, the number must be greater than 54.

$\frac{1}{9}x > 6 \to$ multiply both sides of the inequality by 9, then: $x > 54$

30) Answer: D.

Whole numbers are the basic counting numbers 0, 1, 2, 3, 4, 5, 6, ... and so on.

$3 \times (M + N) = 27$, then $M + N = 9$. $M \geq 0 \to N$ could not be greater than 9

31) Answer: C.

The ratio of lions to tigers is 28 to 12 or 7 to 3 at the zoo. Therefore, total number of lions and tigers must be divisible by 10. $7 + 3 = 10$

From the numbers provided, only 95 is not divisible by 10.

32) Answer: C.

A represents digit 8 in the multiplication.

$5{,}348 \div 14 = 382$

33) Answer: D.

N is even. Let's chooses 2 and 4 for N. Now, let's review the choices provided.

A) $5N = 5 \times 2 = 10$, $5 \times 4 = 20$,   Both results are even.

B) $N + 10 = 2 + 10 = 12$, $4 + 10 = 14$ Both results are even.

C) $\frac{N}{2} = \frac{2}{2} = 1$, $\frac{N}{2} = \frac{4}{2} = 2$ One result is odd and the other one is even.

D) $N + 3 = 2 + 3 = 5$, $4 + 3 = 7$ Both results are odd.

ISEE Lower-Level Subject Test Mathematics

34) Answer: D.

There are equal numbers of four types of cards. Therefore, the total number of cards must be divisible by 4. Only choice D (92) is divisible by 4.

35) Answer: D.

20 seconds is one third of a minute. one third of 75 is 25.

$75 \div 3 = 25$; Jim types 25 words in 20 seconds.

36) Answer: D.

There are equals.

From the choice provided, only choice D is not equal to $\frac{5}{9}$. $\frac{15}{36} = \frac{5}{12}$

37) Answer: C.

The closest to 7.01 is 7 in the choices provided.

38) Answer: C.

Write the numbers in order: 3, 5, 11, 15, 18, 20, 25.

Median is the number in the middle. Therefore, the median is 15.

Practice Tests 1:
Mathematics Achievement

1) Answer: A.

6,682.3749 rounded to the nearest tenth is 6,688.40.

2) Answer: C.

One method to compare fractions is to convert them to decimals.

A. $\frac{2}{7} = 0.285$

B. $\frac{1}{5} = 0.2$

C. $\frac{8}{10} = 0.8$

D. $\frac{6}{10} = 0.6$

0.8 or $\frac{8}{10}$ is the largest number.

3) Answer: D.

If 15 balls are removed from the bag at random, there will be one ball in the bag. The probability of choosing a brown ball is 1 out of 16. Therefore, the probability of not choosing a brown ball is 15 out of 16 and the probability of having not a brown ball after removing 15 balls is the same.

4) Answer: B.

The question is this: 1.89 is what percent of 1.35?

Use percent formula: part = $\frac{percent}{100}$ × whole

part = $\frac{percent}{100}$ × 1.35 ⇒ 1.89 = $\frac{percent \times 1.35}{100}$ ⇒ 189 = percent × 1.35

⇒ percent = $\frac{189}{1.35}$ = 140

5) Answer: D.

The question is this: 196 is what percent of 280?

Use percent formula: part = $\frac{percent}{100}$ × whole

196 = $\frac{percent}{100}$ × 280 ⇒ 196 × 100 = percent × 280 ⇒

ISEE Lower-Level Subject Test Mathematics

$19,600 =$ percent $\times 280 \Rightarrow$ percent $= \frac{19,600}{280} = 70$

196 is 70 % of 280. Therefore, the discount is: 100% − 70% = 30%

6) Answer: C.

Area of a triangle $= \frac{1}{2}$ (base)(height) $\Rightarrow A = \frac{1}{2}(24)(15) = 180$

7) Answer: A.

All angles in a triangle add up to 180 degrees.

$46 + 87 = 133 \to 180 - 133 = 47$

8) Answer: D.

Find common denominator and solve. $\frac{1}{2} + \frac{2}{9} = \frac{9}{18} + \frac{4}{18} = \frac{13}{18}$

9) Answer: D.

least common multiple (LCM) of 8 and 17 is the smallest number that is divisible by both 8 and 17. LCM = 136

10) Answer: D.

5 hours = 300 minutes. Write a proportion and solve.

$\frac{150}{1} = \frac{300}{x} \to x = \frac{300}{150} = 2$

11) Answer: A.

Let x be the number of raptors Daniel saw on Monday. Then:

$Mean = \frac{x+17+20+15+8}{5} = 15 \to x + 60 = 75 \to x = 75 - 60 = 15$

12) Answer: C.

The factors of 32 are: {1, 2, 4, 8, 16, 32}

18 is not a factor of 32.

13) Answer: A.

Perimeter $= 2(width + length)$; $A = width \times length$

First, find the length of the rectangle.

Perimeter $= 2(width + length) \to 52 = 2(9 + length) \to 52 = 18 + 2(length) \to 34 = 2(length) \to length = 17$

$A = 9 \times 17 = 153$

ISEE Lower-Level Subject Test Mathematics

14) Answer: A.

An obtuse angle is an angle of greater than 90 degrees and less than 180 degrees.

Only choice a is an obtuse angle.

15) Answer: D.

There are 9 squares and 3 of them are shaded. Therefore, 3 out of 9 or $\frac{3}{9} = \frac{1}{3}$ are shaded.

16) Answer: B.

Write a proportion and solve. $\frac{2}{3} = \frac{x}{72}$

Use cross multiplication: $3x = 144 \rightarrow x = 48$

17) Answer: D.

To find the discount, multiply the number by (100% − rate of discount).

Therefore, for the first discount we get: $(700)(100\% - 22\%) = (700)(0.78)$

For the next 13 % discount: $(700)(0.78)(0.87)$

18) Answer: D.

A quadrilateral with one pair of parallel sides is a trapezoid.

19) Answer: A.

Plug in 80 for A in the equation. Only choice A works.

$2A + 40 = 120 + A \rightarrow 2 \times 80 + 40 = 120 + 80 \rightarrow 200 = 200$

20) Answer: C.

1 $hour$: \$4.80. 10 $hours$: $10 \times \$4.80 = \$48$

21) Answer: A.

There are 60 students in the class. 18 of the are male and 42 of them are female.

42 out of 60 are female. Then:

$\frac{42}{60} = \frac{x}{100} \rightarrow 4{,}200 = 60x \rightarrow x = 4{,}200 \div 60 = 70\%$

22) Answer: B.

1 pizza has 8 slices. 12 pizzas contain (12×8) 96 slices.

ISEE Lower-Level Subject Test Mathematics

23) Answer: D.

Since Mike running at 5.5 miles per hour and Julia is running at the speed of 9 miles per hour, each hour their distance decreases by 3.5 miles (9 − 5.5 = 3.5). So, it takes 4 hours to cover distance of 21 miles. 21 ÷ 3.5 = 4

24) Answer: C.

$0.079 \times 100 = 7.9\%$

25) Answer: B.

Since Julie gives 6 pieces of candy to each of her friends, then, then number of pieces of candies must be divisible by 6.

A. 74 ÷ 6 = 12.33

B. 336 ÷ 6 = 56

C. 188 ÷ 6 = 31.33

D. 227 ÷ 6 = 37.83

Only choice B gives a whole number.

26) Answer: A.

$7 × 10 = $70

Petrol use: 10 × 3 = 30 liters

Petrol cost: 30 × $1.4 = $42

Money earned: $70 − $42 = $28

27) Answer: B.

$\frac{3}{1,000} = 0.003$

28) Answer: C.

   23 hr. 10 min.
 − 15 hr. 38 min.
    7hr. 32min.

29) Answer: D.

The failing rate is 48 out of 160 = $\frac{48}{160}$

Change the fraction to percent: $\frac{48}{160} \times 100\% = 30\%$

30 percent of students failed. Therefore, 70 percent of students passed the exam.

WWW.MathNotion.Com

ISEE Lower-Level Subject Test Mathematics

30) Answer: D.

If the length of the box is 24, then the width of the box is one third of it, 8, and the height of the box is 1 (one eighth of the width). The volume of the box is:

V=(length)(width)(height) = $(24)(8)(1) = 192$ cm^3

ISEE Lower-Level Subject Test Mathematics

Practice Tests 2:
Quantitative Reasoning

1) Answer: B.

$10 + 2 = 12$, $12 + 3 = 15$, $15 + 4 = 19$, $19 + 5 = 24$

2) Answer: C.

The length of the rectangle is 28. Then, its width is 4. $28 \div 7 = 4$

Perimeter of a rectangle $= 2 \times width + 2 \times length = 2 \times 4 + 2 \times 28$

$P = 8 + 56 = 64$

3) Answer: C.

$Mary's\ Money = y$; $John's\ Money = y + 17$

$John\ gives\ Mary\ \$30 \rightarrow y + 17 - 30 = y - 13$

4) Answer: D.

Dividing 193 by 6 leaves a remainder of 1.

5) Answer: C.

$9,000 + A - 4,700 = 8,200 \rightarrow 4,300 + A = 8,200$

$\rightarrow A = 8,200 - 4,300 = 3,900$

6) Answer: D.

52 divided by 3, the remainder is 1. Then, 25 divided by 6, the remainder is also 1.

7) Answer: A.

$270 off is the same as 54 percent off. Thus, 54 percent of a number is 270.

Then: $54\%\ of\ x = 270 \rightarrow 0.54x = 270 \rightarrow x = \frac{270}{0.54} = 500$

8) Answer: A.

$\frac{9}{5} = 1.80 > 1.69$, the only choice provided that is less than 1.80 is choice A.

9) Answer: D.

$x + 7 = 12 \rightarrow x + 7 - 7 = 12 - 7 \rightarrow x = 5$

$7y = 63 \rightarrow y = 9$

$y + x = 9 + 5 = 14$

WWW.MathNotion.Com

ISEE Lower-Level Subject Test Mathematics

10) Answer: C.

$638 - x + 250 = 518 \rightarrow 888 - x = 518 \rightarrow 888 = 518 + x$

$\rightarrow x = 888 - 518 = 370$

11) Answer: C.

All angles in a triangle sum up to 180 degrees. The triangle provided is an isosceles triangle. In an isosceles triangle, the three angles are 45, 45, and 90 degrees. Therefore, the value of x is 45.

12) Answer: D.

50 percent of $64.00 is $32. (Remember that 50 percent is equal to half)

13) Answer: C.

$14.58 - 9.6 = 4.98$

14) Answer: A.

$\frac{17}{9} - \frac{8}{9} = \frac{9}{9} = 1$

15) Answer: C.

Seven times a number N is $7 \times N$. When 5 is added to it, the result is:

$5 + (7 \times N) = 52 \rightarrow 7N + 5 = 52$

16) Answer: B.

$82 = 5 \times Z + 17 \rightarrow 5Z = 82 - 17 \rightarrow 5Z = 65 \rightarrow Z = 13$

17) Answer: A.

$3,700 - 859 = 2,841$

18) Answer: B.

4 percent of $650 = \frac{4}{100} \times 650 = \frac{1}{25} \times 650 = \frac{650}{25} = 26$

19) Answer: D.

$75 \div (15 + 10) = 75 \div 25 = 3$ not 12

20) Answer: D.

Speed $= \frac{distance}{time} \rightarrow 79 = \frac{4,200}{time} \rightarrow time = \frac{4,200}{79} = 53.16$

It takes Nicole about 53 hours to go from city A to city B.

WWW.MathNotion.Com

ISEE Lower-Level Subject Test Mathematics

21) Answer: C.

$\frac{56}{70} = 0.8$

22) Answer: D.

4 is NOT a prime factor.

23) Answer: B.

$240 \div 60 = 4$

24) Answer: B.

Let's order number of shoes sold per month:

$35, 38, 40, 50, 55, 55$

Median is the number in the middle. Since there are 6 numbers (an even number) the Median is the average of numbers 3 and 4: Median is: $\frac{40+50}{2} = 45$

25) Answer: B.

$390 + 47 = 437$

26) Answer: A.

$11a + 30 = 151$

$11a = 151 - 30$

$11a = 121 \Rightarrow a = 11$

27) Answer: B.

Thousandths

28) Answer: C.

$\frac{95}{100,000} = 0.00095$

29) Answer: C.

$\frac{1}{8}$ of 24 hours is 8 hours. $\frac{1}{8} \times 24 = \frac{24}{8} = 3$

30) Answer: C.

The ratio of red marbles to blue marbles is 3 to 2. Therefore, the total number of marbles must be divisible by 5. $3 + 2 = 5$

48 is the only one that is not divisible by 5.

ISEE Lower-Level Subject Test Mathematics

31) Answer: B.

Area of a square = $side \times side = 16 \rightarrow side = 4$

Perimeter of a square = $4 \times side = 4 \times 4 = 16$

32) Answer: D.

$300 + \square - 220 = 1{,}750 \rightarrow 80 + \square = 1{,}750$

$\square = 1{,}750 - 80 = 1{,}670$

33) Answer: C.

$\frac{1}{5}$ of students are girls. Therefore, $\frac{4}{5}$ of students in the class are boys. $\frac{4}{5}$ of 85 is 68.

There are 68 boys in the class. $\frac{4}{5} \times 85 = \frac{340}{5} = 68$

34) Answer: A.

$N \times (7 - 4) = 48 \rightarrow N \times 3 = 48 \rightarrow N = 16$

35) Answer: B.

If $x \blacksquare y = 2x + 5y - 12$, Then:

$2 \blacksquare 6 = 2(2) + 5(6) - 12 = 4 + 30 - 12 = 22$

36) Answer: C.

Of the numbers provided, 0.7872 is the greatest.

37) Answer: A.

$\frac{7}{8} - \frac{3}{4} = \frac{7}{8} - \frac{6}{8} = \frac{1}{8} = 0.125$

38) Answer: B.

The closest number to 4.03 is 4.

ISEE Lower-Level Subject Test Mathematics

Practice Tests 2:
Mathematics Achievement

1) Answer: C.

The factors of 16 are: {1, 2, 4, 8, 16}

The factors of 38 are: {1, 2, 19, 38}

greatest common factor (GCF) = 2

2) Answer: A.

Sixty-nine thousand, two hundred fourteen is written as 69,214.

3) Answer: D.

The name of a rectangle with sides of equal length is square.

4) Answer: C.

The question is that number 35,648.52 is how many times of number 3.564852.

The answer is 10,000.

5) Answer: B.

Use the Pythagorean Theorem to find the length of the third side:

$a^2 + b^2 = c^2 \Rightarrow 12^2 + 9^2 = c^2 \Rightarrow 225 = c^2 \Rightarrow c = 15$

The perimeter of the triangle is: $15 + 12 + 9 = 36$

6) Answer: A.

Simplify each choice provided using order of operations rules.

A. $10 - (-2) + (-25) = 10 + 2 - 25 = -13$

B. $-8 + (-5) \times (-2) = -8 + 10 = 2$

C. $(-4) \times (-6) + (-2) \times (-7) = 24 + 14 = 38$

D. $(-5) \times (-9) + 20 = 45 + 20 = 65$

Only choice A is -13.

7) Answer: C.

Lily eats 9 pancakes in 1 minute $\Rightarrow$ Lily eats 9×4 pancakes in 4 minutes.

Ella eats $4\frac{1}{2}$ pancakes in 1 minute $\Rightarrow$ Ella eats $4\frac{1}{2} \times 4$ pancakes in 4 minutes.

In total Lily and Ella eat $36 + 18 = 54$ pancakes in 4 minutes.

WWW.MathNotion.Com

ISEE Lower-Level Subject Test Mathematics

8) Answer: C.

$0.72 + 2.2 + 3.68 = 6.6$

9) Answer: B.

Perimeter of a rectangle $= 2(length + width) = 2(7 + 5) = 24$

10) Answer: D.

To solve this problem, divide $4\frac{2}{5}$ by $\frac{1}{10}$.

$4\frac{2}{5} \div \frac{1}{10} = \frac{22}{5} \div \frac{1}{10} = \frac{22}{5} \times \frac{10}{1} = 44$

11) Answer: B.

$\frac{7}{15}$ means 7 is divided by 15. The fraction line simply means division or $\div$.

Therefore, we can write $\frac{7}{15}$ as $7 \div 15$.

12) Answer: C.

Subtract hours: $5 - 6 = -1$

Subtract the minutes: $15 - 30 = -15$

The minutes are less than 0, so:

Add 60 to minutes ($-15 + 60 = 45$ minutes)

Subtract 1 from hours ($-1 - 1 = -2$) the hours are less than 0, add 12: ($12 - 2 = 10$)

The answer is 22:45 that is equal to 10:45

13) Answer: A.

From choices provided, only 116 is NOT a multiple of 6.

14) Answer: C.

Area $= width \times height$

Area $= 84$, $Width = 7$. $\Rightarrow 84 = 7 \times height$

height $= \frac{84}{7} = 12$

15) Answer: B.

Low temperature is 47°f cooler than the temperature at 12:00 PM that is 79°f, that means low temperature is 32°f (79°f – 47°f).

16) Answer: B.

$(8^2 - 5^2) \div (3^3 \div 3^2) = (64 - 25) \div (27 \div 9) = (39) \div (3) = 13$

ISEE Lower-Level Subject Test Mathematics

17) Answer: C.

$19{,}640 \le 22{,}856 \le 26{,}753 \le 26{,}821$

18) Answer: C.

$6.18 + 8.13 + 4.20 + 6.15 + 9.95 + 11.85 =$

$6 + 8 + 4 + 6 + 10 + 12 = 46$

19) Answer: B.

The line is from 6 to -4. $6 - (-4) = 6 + 4 = 10$

20) Answer: D.

Find the least common denominator (LCD), then rewriting each term as an equivalent fraction with the LCD.

Then we compare the numerators of each fraction and put them in correct order from least to greatest or greatest to least.

LCD of 6, 8, 4 and 24 is 24. Rewrite the input fractions as equivalent fractions using the LCD:

A. $\dfrac{20}{24}$ B. $\dfrac{9}{24}$ C. $\dfrac{18}{24}$ D. $\dfrac{7}{24}$

So, Answer: D has the least value.

21) Answer: C.

The first picture is divided to 12 parts that 6 parts of it is shaded ($\dfrac{6}{12} = \dfrac{1}{2}$). The second picture is divided to 8 parts that 3 part of that is shaded ($\dfrac{3}{8}$).

22) Answer: C.

The digit 7 has a value of $7 \times \dfrac{1}{10}$

The digit 5 has a value of $5 \times \dfrac{1}{100}$

The digit 9 has a value of $9 \times 100 = 900$

The digit 1 has a value of 1×10

23) Answer: B.

Three – digit even numbers that have a 6 in the hundreds place and a 9 in the tens place are 690, 692, 694, 696, 698. 694 is one of the choices.

ISEE Lower-Level Subject Test Mathematics

24) Answer: C.

0.55 is equal to $\frac{55}{100} = \frac{11}{20}$

25) Answer: B.

average (mean) = $\frac{sum\ of\ terms}{number\ of\ terms} = \frac{6+10+16+18+16+18+19.6}{7} = 14.8$

26) Answer: A.

probability = $\frac{desired\ outcomes}{possible\ outcomes} = \frac{5}{5+6+11+10} = \frac{5}{32}$

27) Answer: A.

To find total number of students in Riddle Elementary school, add number of all students. $66 + 17 = 83$

28) Answer: D.

$1\ year = 365\ days, 1\ day = 24\ hours$

$1\ year = 365 \times 24$

$1\ year = 8,760 \Rightarrow 3\ year = 3 \times 8,760 = 26,280$

29) Answer: A.

$\frac{19}{31} \cong 0.61$

30) Answer: B.

$30 \times 18 = 30 \times (10 + 8) = 30 \times 10 + 30 \times 8$

"End"

www.ingramcontent.com/pod-product-compliance
Lightning Source LLC
Chambersburg PA
CBHW080438110426
42743CB00016B/3208